PoliTO Springer Series

Springer, in cooperation with Politecnico di Torino, publishes the PoliTO Springer Series. This co-branded series of publications includes works by authors and volume editors mainly affiliated with Politecnico di Torino and covers academic and professional topics in the following areas: Mathematics and Statistics, Chemistry and Physical Sciences, Computer Science, All fields of Engineering. Interdisciplinary contributions combining the above areas are also welcome. The series will consist of lecture notes, research monographs, and briefs. Lectures notes are meant to provide quick information on research advances and may be based e.g. on summer schools or intensive courses on topics of current research, while SpringerBriefs are intended as concise summaries of cutting-edge research and its practical applications. The PoliTO Springer Series will promote international authorship, and addresses a global readership of scholars, students, researchers, professionals and policymakers.

More information about this series at http://www.springer.com/series/13890

Fernando Corinto · Alessandro Torcini
Editors

Nonlinear Dynamics in Computational Neuroscience

POLITECNICO
DI TORINO

Springer

Editors
Fernando Corinto
Dipartimento di Elettronica e Telecomunicazioni (DET)
Politecnico di Torino
Turin
Italy

Alessandro Torcini
Laboratoire de Physique Théorique et Modélisation
Université de Cergy-Pontoise
Cergy-Pontoise
France

ISSN 2509-6796 ISSN 2509-7024 (electronic)
PoliTO Springer Series
ISBN 978-3-030-10009-4 ISBN 978-3-319-71048-8 (eBook)
https://doi.org/10.1007/978-3-319-71048-8

Printed on acid-free paper

This Springer imprint is published by the registered company Springer International Publishing AG part of Springer Nature
The registered company address is: Gewerbestrasse 11, 6330 Cham, Switzerland

Foreword

This book is devoted to the theme of *Nonlinear Dynamics in Computational Neuroscience*, a very dynamic and interdisciplinary area of current research. It combines a set of unique contributions from experts in neuroscience, computer science, physics, mathematics, and engineering who attended the 2015 International Workshop on *Nonlinear Dynamics in Computational Neuroscience: from Physics and Biology to ICT* organized by SICC, the Italian Society for Chaos and Complexity (http://www.sicc-it.org).

As the current President of the Society, it is my pleasure to present this book to the reader as a collection of insightful chapters on different aspects of this interesting and promising research field. The program of the workshop as well as the contents of this book was organized and edited by Alessandro Torcini and Fernando Corinto, both members of our society. I wish to thank for both organizing such a successful event and bringing together the unique set of experts that contributed to this volume.

I am sure that the reader will find this book enjoyable and motivating. Computational neuroscience is an exciting emerging research area where the interdisciplinary cross-fertilization from different areas of science and engineering is required. Complexity science and nonlinear dynamics can offer invaluable tools and approaches to tackle the many challenging open problems in this area. For this reason, I believe this book shows the importance of interdisciplinary research, whose promotion is at the core of the activities of the Italian Society for Chaos and Complexity and its members.

Napoli, Italy
Bristol, UK

Mario Di Bernardo
Department of Electrical Engineering and Information Technology, University of Naples Federico II

and

Department of Engineering Mathematics
University of Bristol

Contents

Introduction

This Lecture Note on *Nonlinear Dynamics in Computational Neuroscience* collects researchers' contributions working across (and between) disciplines linked to Computational Neuroscience. In particular, it summarizes the most recent results presented at the Workshop Nonlinear Dynamics in Computational Neuroscience: from Physics and Biology to ICT, held at Valentino Castle, Polytechnic University of Turin, Italy in September 2015. The workshop gathered more than 50 researchers and Ph.D. students coming from different communities: namely, computational neuroscientists, physicists, neurophysiologists and neural engineers. The content of this volume ranges from nonlinear dynamical analysis to the understanding of neural computation to physiology, from the simulations of brain circuits to the development of engineerized devices and platforms for neuromorphic computation.

The volume contains eight chapters encompassing relevant issues in complex neural modeling, ranging from mean field models to spiking neuron models, from memristor based neural networks to circuit modeling and neural signals analysis. In particular, the book opens with a chapter devoted to a new and interesting analytic approach able to give an exact mesoscopic modelization of microscopic spiking neural networks in terms of the average firing rate and membrane potential [9, 11]. This methodology has been developed for instantaneous synapses in [9], S. Coombes and A. Byrne in their article extend it to more realistic synaptic transmissions paving the way for the development of the next generation of *neural mass models* [5]. The contribution of M. Mattia and M. V. Sanchez-Vives reported in the second chapter is a stimulating review on mesoscopic models developed for cortical modules and capable to reproduce slow oscillations [13] measured at the various scales in the brain. How to deal with the treatment of the neural signals obtained at different scales is the topic of the next chapter. Specifically, F. Vallone, M. Caleo and A. Di Garbo, review nonlinear time series analysis methods [8] developed to characterize neural signals.

The following two chapters are devoted to the control and characterization of collective oscillations emerging in neural networks with different topologies. In particular, chapter four by S. Luccioli, A. Barzilai, E. Ben-Jacob, P. Bonifazi, and A. Torcini reviews recent computational results concerning the possibility to orchestrate the dynamics of an entire network by acting on a single neuron, these analyses were stimulated by recent experimental findings reported for the hyppocampus in its first stages of developments [3]. The following chapter by S. Olmi and A. Torcini addresses a quite active topic of research in nonlinear dynamics, namely the emergence of states with broken symmetry, *Chimera States*, in symmetrically coupled networks [1, 2, 10]. In particular, the authors examine the spontaneous emergence of various kind of chimeras and their stability in networks of pulse coupled neurons with respect to noise and random dilution.

Chapter 6, authored by A. Aloisi, D. Pisignano, and R. Rinaldi, report a detailed description of innovative applications of nanothecnology in neuroscience, in particular discussing three examples lying at the border between research and medical applications. The last two chapters are devoted to memristor-based neurons and synapses [4, 12], which are fundamental elements for the development of neuromorphyc circuits [6, 7]. In particular, chapter seven by A. Ascoli, R. Tetzlaff, and M. Biey deals with a nonlinear system theory-based approach to determine circuit quantities to a high degree of accuracy for neuromorphyc circuits. Finally, a memristor-based cell model is introduced and analyzed by A. Buscarino, C. Corradino, L. Fortuna, M. Frasca, and V.-T. Pham, this model is capable to reproduce different dynamical behaviours, ranging form wave propagation to the emergence of Turing patterns.

Fernando Corinto
Dipartimento di Elettronica e Telecomunicazioni (DET)
Politecnico di Torino, Turin, Italy

Alessandro Torcini
Laboratoire de Physique Théorique et Modélisation
Université de Cergy-Pontoise - CNRS
UMR 8089, Cergy-Pontoise, France

and

Inserm, INMED, Institute de Neurobiologie de la
Méditerranée and INS, Institut de Neurosciences des Systémes
Aix-Marseille Université, Marseille, France

and

CNRS, CPT, UMR 7332
Aix-Marseille Université, Université de Toulon
Marseille, France

and

CNR - Consiglio Nazionale delle Ricerche - Istituto
dei Sistemi Complessi, Sesto Fiorentino, Italy

References

1. Abrams, D.M., Strogatz, S.H.: Phys. Rev. Lett. **93**, 174102 (2004)
2. Abrams, D.M., Mirollo, R.E., Strogatz, S.H., Wiley, D.A.: Phys. Rev. Lett. **101**, 084103 (2008)
3. Bonifazi, P., Goldin, M., Picardo, M.A., Jorquera, I., Cattani, A., et al.: Science **326**, 1419–1424 (2009)
4. Chua, L.: IEEE Trans. Circuit Theory **18**, 507–519 (1971)
5. Deco, G., et al.: PLoS Comput. Biol. **4**, e1000092 (2008)
6. Indiveri, G., et al.: Nanotechnology **24**, 384010 (2013)
7. Jo, S.H., et al.: Nano Lett. **10**, 1297–1301 (2010)
8. Kantz, H., Schreiber, T.: Nonlinear Time Series Analysis, vol. 7. Cambridge University Press, Cambridge (2004)
9. Montbrió, E., Pazó, D., Roxin, A.: Phys. Rev. X **5**, 021028 (2015)
10. Panaggio, M.K., Abrams, D.M.: Nonlinearity **28**, R67 (2015)
11. Pazó, D., Montbrió, E.: Phys. Rev. X **4**, 011009 (2014)
12. Strukov, D.B., Snider, G.S., Stewart, D.R., Williams, R.S.: Nature **453**, 80 (2008)
13. Van Someren, E.J.: Slow Brain Oscillations of Sleep, Resting State and Vigilance, vol. 193. Elsevier, Amsterdam (2011)

Next Generation Neural Mass Models

Stephen Coombes and Áine Byrne

Abstract Neural mass models have been actively used since the 1970s to model the coarse grained activity of large populations of neurons and synapses. They have proven especially useful in understanding brain rhythms. However, although motivated by neurobiological considerations they are phenomenological in nature, and cannot hope to recreate some of the rich repertoire of responses seen in real neuronal tissue. In this chapter we consider the θ-neuron model that has recently been shown to admit to an exact mean-field description for instantaneous pulsatile interactions. We show that the inclusion of a more realistic synapse model leads to a mean-field model that has many of the features of a neural mass model coupled to a further dynamical equation that describes the evolution of network synchrony. A bifurcation analysis is used to uncover the primary mechanism for generating oscillations at the single and two population level. Numerical simulations also show that the phenomena of event related synchronisation and desynchronisation are easily realised. Importantly unlike its phenomenological counterpart this *next generation neural mass model* is an exact macroscopic description of an underlying microscopic spiking neurodynamics, and is a natural candidate for use in future large scale human brain simulations.

1 Introduction

The term *neural mass model* is often used to refer to low dimensional models that aim to describe the coarse grained activity of large populations of neurons and synapses. They are typically cast as systems of ordinary differential equations (ODEs) and in their modern incarnations are exemplified by variants of the two dimensional Wilson–Cowan model [1]. This model tracks the activity of an excitatory population of neurons coupled to an inhibitory population. With the augmentation of such models by more realistic forms of synaptic and network interaction they have proved especially successful in providing fits to neuroimaging data. Historically one of the

S. Coombes (✉) · Á. Byrne
School of Mathematical Sciences, Centre for Mathematical Medicine and Biology,
University of Nottingham, University Park, Nottingham NG7 2RD, UK
e-mail: stephen.coombes@nottingham.ac.uk

F. Corinto and A. Torcini (eds.), *Nonlinear Dynamics in Computational Neuroscience*, PoliTO Springer Series, https://doi.org/10.1007/978-3-319-71048-8_1

first examples in this area is the Zetterberg model [2] for the electroencephalogram (EEG) rhythm. This is based on previous ideas developed by Lopes da Silva and colleagues [3, 4] and is built from three interacting neural mass models, as a minimal model of a cortical column. The first represents a population of pyramidal cells, the second a population of excitatory interneurons, and the third a population of inhibitory interneurons. Since its introduction the Zetterberg model has become more widely popularised by the work of Jansen and Rit [5] and used to explain epileptic brain dynamics, particularly by Wendling and colleagues, as recently reviewed in [6]. Another well known neural mass model is that of Liley, which pays particular attention to the role of synaptic reversal potentials, and see [7] for a discussion of this model within the context of Freeman's ideas on the importance of chaos for cognition and perception [8]. As well as proving useful for understanding EEG rhythms ranging from delta (1–4 Hz) through to gamma (30–70 Hz) [9], neural mass models have been used to describe brain resonance phenomena [10], resting brain state activity [11] and are very popular in the neuroimaging community. In this latter instance they are often used for model driven fusion of multiple neuroimaging modalities, such as EEG and functional magnetic resonance imaging (fMRI) [12], as well as to augment the dynamic causal modelling framework for understanding how event-related responses result from the dynamics of coupled neural populations [13]. Moreover, they are now an integral part of the Virtual Brain project that aims to deliver the first open simulation of the human brain based on individual large-scale connectivity [14], as well as play a key role in the neuro-computational modelling of neurological and psychiatric disorders [15]. This latter work is especially viable since neural mass models can incorporate accurate descriptions of synaptic processing, typically in the form of a synaptic response function that is driven by firing *rate* rather than by the arrival times of individual action potentials. However, it is important to remember that at heart all neural mass models to date are essentially phenomenological, with state variables that track coarse grained measures of the average membrane potential, population firing rate or synaptic activity. At best they are expected to provide appropriate levels of description for many thousands of near identical interconnected neurons with a preference to operate in *synchrony*. This latter assumption is especially important for the generation of a sufficiently strong physiological signal that can be detected non-invasively. The variation of synchrony within a neuronal population is believed to underly the decrease or increase of power seen in given EEG frequency bands. The former phenomenon is called event-related desynchronisation (ERD), and the latter event-related synchronisation (ERS) [16]. Unfortunately the assumption of synchrony within neural mass models means that they cannot hope to describe ERD and ERS, at least not at the single population level. Rather this sets a natural challenge for the next generation of neural mass models. It is precisely this issue that we take up in this chapter.

As a starting point to move beyond the current neural mass models we draw inspiration from the physics of self-organised networks. In particular the observation of macroscopic coherent states in large networks of coupled spiking neuron models has inspired a search for equivalent low-dimensional dynamical descriptions, and see [17] for a recent review of oscillatory network dynamics in neuroscience.

However, although the mathematical step from microscopic to macroscopic dynamics has proved elusive for the majority of spiking models the θ-neuron model has proven amenable to such a reduction for pulsatile coupling by Luke et al. [18]. A similar approach by Montbrió et al. [19] has been used to reduce networks of quadratic integrate-and-fire neurons. Here we show how to naturally augment these approaches to incorporate the biologically realistic forms of synaptic coupling that are commonly adopted within current neural mass models. In this way we arrive at the first instance of a next generation neural mass model, with a derived (as opposed to postulated) population firing rate that couples directly to a dynamical variable describing population synchrony. In Sect. 2 we discuss the main elements of synaptic processing that are incorporated within standard neural mass models, and give a heuristic description of how to close the equations of motion in terms of population firing rates. The same model of a synapse is used in Sect. 3, though this time driven by the spike times arising in a network of θ-neurons. For a large globally coupled network an exact mean field description is derived, and the form of the equations compared and contrasted with standard phenomenological neural mass models. In Sect. 4 we present a bifurcation analysis for the single and two population mean-field models, and use this to highlight the primary mechanisms for generating population oscillations. Importantly we also show, through direct numerical simulations, that the model supports ERD and ERS. Finally in Sect. 5 we reflect upon the use of such models in future large scale human brain simulations, as well as their subsequent mathematical analysis.

2 Neural Mass Modelling

Neural mass models generate brain rhythms using the notion of population firing rates, aiming to side-step the need for large scale simulations of more realistic networks of spiking neurons. However, both approaches often make use of the same level of description for synaptic processing, in a manner that we shall now clarify.

At a synapse, presynaptic firing results in the release of neurotransmitters that causes a change in the membrane conductance of the post-synaptic neuron. This post-synaptic current may be written $I = g(v_{\text{syn}} - v)$, where v is the voltage of the post-synaptic neuron, v_{syn} is its membrane reversal potential and g is a conductance. This conductance is proportional to the probability that a synaptic receptor channel is in an open conducting state. This probability depends on the presence and concentration of neurotransmitter released by the presynaptic neuron. The sign of v_{syn} relative to the resting potential (assumed to be zero) determines whether the synapse is excitatory ($v_{\text{syn}} > 0$) or inhibitory ($v_{\text{syn}} < 0$). The effect of some synapses can be described with a function that fits the shape of the post-synaptic response due to the arrival of action potential at the pre-synaptic release site. A post-synaptic conductance change $g(t)$ would then be given by $g(t) = ks(t - T)$ for $t \geq T$, where T is the arrival time of a pre-synaptic action potential, $s(t)$ fits the shape of a realistic post-synaptic conductance, and k is a constant. A common (normalised) choice for $s(t)$ is the α-function:

$$s(t) = \alpha^2 t \mathrm{e}^{-\alpha t} \Theta(t), \tag{1}$$

where Θ is a Heaviside step function. The conductance change arising from a train of action potentials, with firing times T^m, is given by

$$g(t) = k \sum_m s(t - T^m). \tag{2}$$

If s is the Green's function of a linear differential operator, so that $Qs = \delta$, then we may write (2) in the equivalent form

$$Qg = k \sum_m \delta(t - T^m). \tag{3}$$

This is indeed the case for the choice (1) for which

$$Q = \left(1 + \frac{1}{\alpha}\frac{\mathrm{d}}{\mathrm{d}t}\right)^2. \tag{4}$$

In many neural population models it is assumed that the interactions are mediated by firing rates rather than action potentials (spikes) *per se*. To see how this might arise we perform a short-time average of (3) over some time-scale τ and assume that s is sufficiently *slow* so that $\langle Qg \rangle_t$ is approximately constant, where

$$\langle x \rangle_t = \frac{1}{\tau}\int_{t-\tau}^{t} x(t')\mathrm{d}t', \tag{5}$$

then we have that $Qg = kf$, where f is the instantaneous firing rate (number of spikes per time Δ). For a single neuron (real or synthetic) experiencing a constant drive it is natural to assume that this firing rate is a function of the drive alone. If for the moment we assume that a neuron spends most of its time close to rest such that $v_{\text{syn}} - v \approx v_{\text{syn}}$, and absorb a factor v_{syn} into k, then for synaptically interacting neurons this drive is directly proportional to the conductance state of the presynaptic neuron. Thus for a single population of identically and globally coupled neurons operating synchronously we are led naturally to equations like:

$$Qg = \kappa f(g), \tag{6}$$

for some strength of coupling κ. A common choice for the *population* firing rate function is the sigmoid

$$f(g) = \frac{f_0}{1 + \mathrm{e}^{-r(g-g_0)}}, \tag{7}$$

which saturates to f_0 for large g. This functional form, with threshold g_0 and steepness parameter r, is not derived from a biophysical model, rather it is seen as a physiologically consistent choice. The extension to multiple interacting populations is straight forward, and the popular Jansen-Rit model [5], provides a classic example of such a generalisation. This can be written in the form

$$Q_E g_P = \kappa_P f(g_E - g_I), \quad Q_E g_E = \kappa_E f(w_1 g_P) + A, \quad Q_I g_I = \kappa_I f(w_2 g_P), \tag{8}$$

which describes a network of interacting pyramidal neurons (P), inhibitory interneurons (I) and excitatory interneurons (E). Here, Q_a is given by (4) under the replacement $\alpha \to \alpha_a$ for $a \in \{E, I\}$, $w_{1,2}$, $\kappa_{E,I,P}$ are constants, and A is an external input. It has been used to model both normal and epileptic patterns of cortical activity and its bifurcation structure has been systematically analysed in [20, 21]. Despite its usefulness in describing certain large scale brain rhythms, and especially alpha (8–13 Hz), it suffers the same deficiencies as all other neural mass models, namely it cannot track the level of synchrony *within* a neuronal population.

3 θ-Neuron Network and Reduction

The θ-neuron model or Ermentrout–Kopell canonical model is now widely known throughout computational neuroscience as a parsimonious model for capturing the firing and response properties of a cortical cell [22]. It is described by a purely one dimensional dynamical system evolving on a circle according to

$$\frac{\mathrm{d}}{\mathrm{d}t}\theta = (1 - \cos\theta) + (1 + \cos\theta)\eta, \qquad \theta \in [-\pi, \pi), \tag{9}$$

where η represents a constant drive. For $\eta < 0$ the θ-neuron supports a pair of equilibria $\theta_\pm$, with $\theta_+ < 0$ and $\theta_- > 0$, and no equilibria for $\eta > 0$. In the former case the equilibria at θ_+ is stable and the one at θ_- unstable. In neurophysiological terms, the unstable fixed point at θ_- is a threshold for the neuron model. Any initial conditions with $\theta \in (\theta_+, \theta_-)$ will be attracted to the stable equilibrium, while initial data with $\theta > \theta_-$ will make a large excursion around the circle before returning to the rest state. For $\eta > 0$ the θ-neuron oscillates with frequency $2\sqrt{\eta}$. When $\eta = 0$ the θ-neuron is poised at a saddle-node on an invariant circle (SNIC) bifurcation.

A network of θ-neurons can be described with the introduction of an index $i = 1, \ldots, N$ and the replacement $\eta \to \eta_i + I_i$, where I_i describes the synaptic input current to neuron i. For a globally coupled network this can be written in the form $I_i = g(t)(v_{\text{syn}} - v_i)$ for some global conductance g and local voltage v_i. As a model for the conductance we take the form used in (3) and write

$$Qg(t) = \frac{k}{N} \sum_{j=1}^{N} \sum_{m \in \mathbb{Z}} \delta(t - T_j^m), \tag{10}$$

where T_j^m is the mth firing time of the jth neuron. These are defined to happen every time θ_j increases through π. It is well known that the θ-neuron model is formally equivalent to a quadratic integrate-and-fire model for voltage dynamics [23] under the transformation $v_i = \tan(\theta_i/2)$ (so that $\cos\theta_i = (1-v_i^2)/(1+v_i^2)$ and $\sin\theta_i = 2v_i/(1+v_i^2)$). This voltage relationship allows us to write the network dynamics as

$$\frac{\mathrm{d}}{\mathrm{d}t}\theta_i = (1-\cos\theta_i) + (1+\cos\theta_i)(\eta_i + g(t)v_{\text{syn}}) - g(t)\sin\theta_i, \tag{11}$$

$$Qg = 2\frac{k}{N}\sum_{j=1}^{N} P(\theta_j). \tag{12}$$

Here $P(\theta) = \delta(\theta - \pi)$ and is periodically extended such that $P(\theta) = P(\theta + 2\pi)$, and we have used the result that $\delta(t - T_j^m) = \delta(\theta_j(t) - \pi)|\dot{\theta}_j(T_j^m)|$. For the case that $v_{\text{syn}} \gg v_i$, $Q = 1$, and $P(\theta)$ has a shape of the form $(1 - \cos(\theta))^n$ for some positive integer n we recover the model of Luke et al. [18]. In this case these authors have shown how to obtain an exact mean-field reduction making use of the Ott–Antonsen (OA) ansatz. The same style of reduction has also been used by Pazó and Montbrió to study pulse-coupled Winfree networks [24]. The OA anstaz was originally used to find solutions on a reduced invariant manifold of the Kuramoto model [25], and essentially assumes that the distribution of phases as $N \to \infty$ has a simple unimodal shape, capable of describing synchronous (peaked) and asynchronous (flat) distributions. In the following we show how their reduction approach extends to the more biologically realistic model described by (11)–(12) that includes synaptic reversal potentials and causal non-instantaneous synaptic responses. We note that even in the limit of fast synaptic interactions we do not recover models of the type described in [18, 19, 24] due to our focus on conductance changes and the inclusion of voltage shunts.

3.1 Mean Field Reduction

In the following we shall choose the background drives η_i to be random variables drawn from a Lorentzian distribution $L(\eta)$ with

$$L(\eta) = \frac{1}{\pi}\frac{\Delta}{(\eta - \eta_0)^2 + \Delta^2}, \tag{13}$$

where η_0 is the centre of the distribution and Δ the width at half maximum. For the choice of Q we shall take Eq. (4). In the coupled network, and if the frequencies of the individual neurons are similar enough, then one may expect some degree of phase locking (ranging from synchrony to asynchrony), itself controlled in part by the the time-to-peak, $1/\alpha$, of the synaptic filter. In the limit $N \to \infty$ the state of the network at time t can be described by a continuous probability distribution function

$\rho(\eta, \theta, t)$, which satisfies the continuity equation (arising from the conservation of oscillators):

$$\frac{\partial}{\partial t}\rho + \frac{\partial}{\partial \theta}\rho c = 0, \tag{14}$$

where c is a given realisation of $\dot{\theta}$,

$$c = (1 - \cos\theta) + (1 + \cos\theta)(\eta + g v_{\text{syn}}) - g \sin\theta. \tag{15}$$

The global drive to the network given by the right hand side of (12) can be constructed as

$$\lim_{N\to\infty} \frac{1}{N}\sum_{j=1}^{N} P(\theta_j) = \int_0^{2\pi} \mathrm{d}\theta \int_{-\infty}^{\infty} \mathrm{d}\rho(\eta, \theta, t) P(\theta). \tag{16}$$

Hence, the evolution of the synaptic conductance g is given by

$$Qg = \frac{k}{\pi}\sum_{m\in\mathbb{Z}} \int_0^{2\pi} \mathrm{d}\theta \int_{-\infty}^{\infty} \mathrm{d}\rho(\eta, \theta, t) \mathrm{e}^{im(\theta-\pi)}, \tag{17}$$

where we have used the result that $2\pi P(\theta) = \sum_{m\in\mathbb{Z}} \mathrm{e}^{im(\theta-\pi)}$. The formula for c above may be written conveniently in terms of $\mathrm{e}^{\pm i\theta}$ as

$$c = f\mathrm{e}^{i\theta} + h + \overline{f}\mathrm{e}^{-i\theta}, \tag{18}$$

where $f = ((\eta - 1) + v_{\text{syn}} g + ig)/2$ and $h = (\eta + 1) + v_{\text{syn}} g$, and $\overline{f}$ denotes the complex conjugate of f.

The OA ansatz assumes that $\rho(\eta, \theta, t)$ has the product structure $\rho(\eta, \theta, t) = L(\eta)F(\eta, \theta, t)$. Since $F(\eta, \theta, t)$ should be 2π periodic in θ it can be written as a Fourier series:

$$F(\eta, \theta, t) = \frac{1}{2\pi}\left\{1 + \sum_{n=1}^{\infty} F_n(\eta, t)\mathrm{e}^{in\theta} + \text{cc}\right\}, \tag{19}$$

where cc denotes complex conjugate. The insight in [26] was to restrict the Fourier coefficients such that $F_n(\eta, t) = a(\eta, t)^n$, where $|a(\eta, t)| \leq 1$ to avoid divergence of the series. There is also a further requirement that $a(\eta, t)$ can be analytically continued from real η into the complex η-plane, and that this continuation has no singularities in the lower half η-plane, and that $|a(\eta, t)| \to 0$ as $\text{Im}\,\eta \to -\infty$. If we now substitute (18) into the continuity equation (14), use the OA ansatz, and balance terms in $\mathrm{e}^{i\theta}$ we obtain an evolution equation for $a(\eta, t)$ as

$$\frac{\partial}{\partial t}a - ia^2 f - iah - i\overline{f} = 0. \tag{20}$$

It is now convenient to introduce the Kuramoto order parameter

$$Z(t) = \int_0^{2\pi} \mathrm{d}\theta \int_{-\infty}^{\infty} \mathrm{d}\eta \rho(\eta, \theta, t) \mathrm{e}^{i\theta}, \tag{21}$$

where $|Z| \leq 1$. Using the OA ansatz (and using the orthogonality properties of $\mathrm{e}^{i\theta}$, namely $\int_0^{2\pi} \mathrm{e}^{ip\theta}\mathrm{e}^{iq\theta}\mathrm{d}\theta = 2\pi\delta_{p+q,0}$) we then find that

$$\overline{Z}(t) = \int_{-\infty}^{\infty} \mathrm{d}\eta L(\eta) a(\eta, t). \tag{22}$$

By noting that the Lorentzian (13) has simple poles at $\eta_\pm = \eta_0 \pm i\Delta$ the integral in (22) may be performed by choosing a large semi-circle contour in the lower half η-plane. This yields $\overline{Z}(t) = a(\eta_-, t)$, giving $Z(t) = a(\eta_+, t)$. Hence, the dynamics for g given by (17) can be written as $Qg = \kappa f(Z)$, where $\kappa = k/\pi$ and

$$f(Z) = 1 + \left[\sum_{m=1}^{\infty} (-Z)^m + \mathrm{cc} \right] = \frac{1 - |Z|^2}{1 + Z + \overline{Z} + |Z|^2}, \qquad |Z| < 1. \tag{23}$$

The dynamics for Z is obtained from (20) as $\mathrm{d}Z/\mathrm{d}t = \mathcal{F}(Z; \eta_0, \Delta) + \mathcal{G}(Z, g; v_{\mathrm{syn}})$, where

$$\mathcal{F}(Z; \eta_0, \Delta) = -i\frac{(Z-1)^2}{2} + \frac{(Z+1)^2}{2}\left[-\Delta + i\eta_0\right], \tag{24}$$

$$\mathcal{G}(Z, g; v_{\mathrm{syn}}) = i\frac{(Z+1)^2}{2} v_{\mathrm{syn}} g - \frac{Z^2 - 1}{2} g. \tag{25}$$

Here we may view (24) as describing the intrinsic population dynamics and (25) the dynamics induced by synaptic coupling. Thus the form of the mean field model is precisely that of a neural mass model as given by Eq. (6). Importantly the firing rate f is a derived quantity that is a real function of the complex Kuramoto order parameter for synchrony. This in turn is described by a complex ODE with parameters from the underlying microscopic model.

To illustrate the validity of the reduction presented above we show a simulation of a network with $N = 500$ θ-neurons and the mean field model in Fig. 1. Here we plot the real and imaginary parts of Z which we write in the form $R\mathrm{e}^{i\Psi}$ for the mean field reduction and calculate as $Z = N^{-1}\sum_{j=1}^{N} \mathrm{e}^{i\theta_j}$ for the finite size network simulation. It is abundantly clear that the two realisations agree very well. If the size of the population in the large scale simulations is reduced then one can more easily see finite size fluctuations as expected.

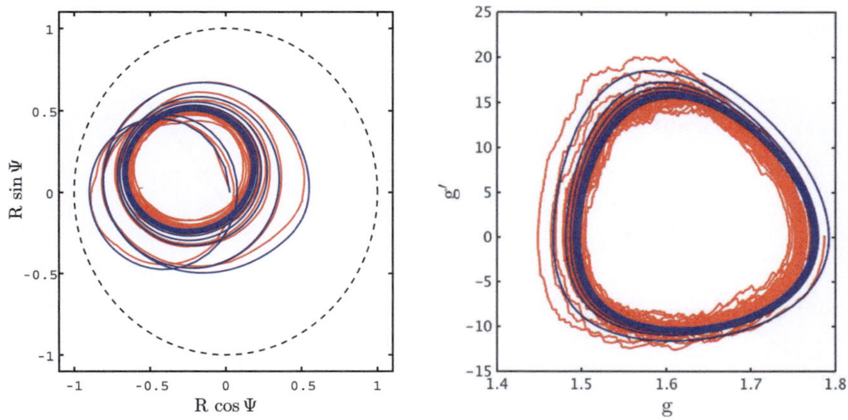

Fig. 1 Comparison between the reduced mean field model (blue) and a simulation of a network of 500 θ-neurons (red). Left: Phase plane projection for the Kuramoto order parameter $Z = Re^{i\Psi}$. Right: Phase plane projection for the synaptic conductance g. Here $\eta_0 = 20$, $\Delta = 0.5$, $v_{\text{syn}} = -10$, $\kappa = 1$, $\alpha = 0.95$

4 Next Generation Neural Mass Model: Analysis

The mean field model derived in Sect. 3.1 is a natural candidate for a next generation neural mass model. It generalises the form of the phenomenological neural mass model whilst maintaining contact with biological reality in that it preserves the notion of both population rate *and* synchrony. An almost identical model has recently been discussed by Laing [27], although here the focus was on a first order synapse model (namely $Q = (1 + \alpha^{-1}\mathrm{d}/\mathrm{d}t)$) with no provision for synaptic reversal potentials. In mathematical terms we are now faced with understanding the dynamics of a coupled system of ODEs given by

$$Qg = \kappa f(Z), \qquad \frac{\mathrm{d}}{\mathrm{d}t} Z = \mathscr{F}(Z; \eta_0, \Delta) + \mathscr{G}(Z, g; v_{\text{syn}}), \tag{26}$$

with f, $\mathscr{F}$ and $\mathscr{G}$ given by (23)–(25) respectively, and Q a linear differential operator such as given by (4). One practical way to assess the emergent behaviour of the model (26) under parameter variation is through numerical bifurcation analysis. We now pursue this for (26), as well as for its natural extension to cover two interacting populations.

4.1 *Bifurcation Diagrams*

We first consider the case of a purely inhibitory population. Using XPPAUT [28] we find that for a wide range of system parameters it is possible to find a Hopf bifurcation of a steady state to a periodic orbit under parameter variation of η_0 (controlling the

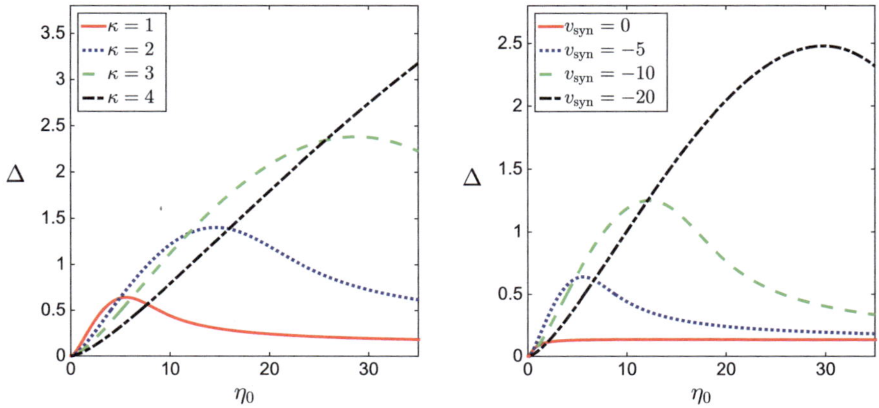

Fig. 2 Two parameter continuation of a Hopf bifurcation in the single population model described by (26) using (4) with $\alpha = 1$. Left: Curves obtained for various values of κ with $v_{\rm syn} = -5$. Right: Curves obtained under for various values of $v_{\rm syn}$ with $\kappa = 1$. In both diagrams the area under the curves represents the parameter window for oscillatory behaviour

mean value of the background drive). To illustrate the relatively large volume of parameter space that can support oscillations by this mechanism we show a two parameter continuation of the Hopf bifurcation in (Δ, η_0) (controlling the shape of the distribution (13)) in Fig. 2, for several values of the coupling strength κ and reversal potential $v_{\rm syn}$. Since the OA ansatz does not hold for $\Delta = 0$ (so that some degree of heterogeneity must be present) and oscillations are only possible for $\eta_0 > 0$, the bifurcation diagrams are only presented for the scenario $\Delta, \eta_0 > 0$.

Suppose now that we have two populations, one excitatory and one inhibitory, with reciprocal connections. Introducing the labels E and I for each population then the natural generalisation of (26) is

$$Q_{ab} g_{ab} = \kappa_{ab} f(Z_b), \qquad \frac{\mathrm{d}}{\mathrm{d}t} Z_a = \mathscr{F}_a(Z_a) + \sum_b \mathscr{G}_b(Z_a, g_{ab}), \tag{27}$$

where $a, b \in \{E, I\}$. Here, Q_{ab} is obtained from (4) under the replacement $\alpha \to \alpha_{ab}$, $\mathscr{F}_a(Z_a) = \mathscr{F}(Z_a; \eta_0^a, \Delta^a)$ and $\mathscr{G}_b(Z_a, g_{ab}) = \mathscr{G}(Z_a, g_{ab}; v_{syn}^{ab})$. The system of equations (27) generalises those recently presented by Laing [27] (to include reversal potentials, higher order synapses and self coupling), who highlighted the ability of such networks to produce a so-called pyramidal-interneuronal network gamma (PING) rhythm [29], as shown in the inset of Fig. 3. Here we also show a bifurcation diagram as a function of κ_{IE}, when $\kappa_{EE} = \kappa_{II} = 0$, which shows that periodic behaviour can be destroyed in a supercritical Hopf bifurcation as κ_{IE} is decreased. In Fig. 4 we show bifurcation diagrams under the variation of κ_{EI} and η_0^I. We see that when κ_{EI} is decreased periodic behaviour can be destroyed in a supercritical Hopf bifurcation. With an increase in η_0^I it is also possible to generate a supercritical Hopf bifurcation to terminate the PING rhythm.

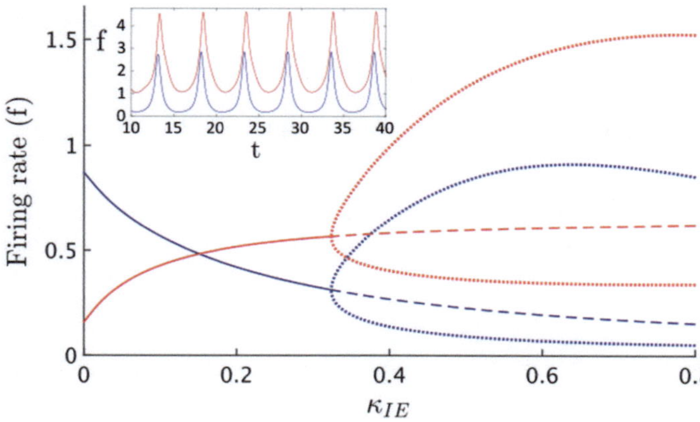

Fig. 3 Bifurcation diagram for the reciprocally connected PING network defined by (27) under variation of κ_{IE}, for both the excitatory (blue) and inhibitory (red) populations. Solid lines: stable; dashed lines: unstable. Circles show maximum and minimum values of $f(Z_E)$ and $f(Z_I)$ over one period of oscillation when no steady states are stable. The inset shows a PING rhythm for $\kappa_{IE} = 0.65$. Parameters: $\alpha_{EI} = 0.8$, $\alpha_{IE} = 10$, $\kappa_{EI} = 0.5$, $v_{\text{syn}}^{EI} = -10$, $v_{\text{syn}}^{IE} = 10$, $\eta_0^E = 10$, $\eta_0^I = 0$, $\Delta^E = \Delta^I = 0.5$, $\kappa_{EE} = \kappa_{II} = 0$

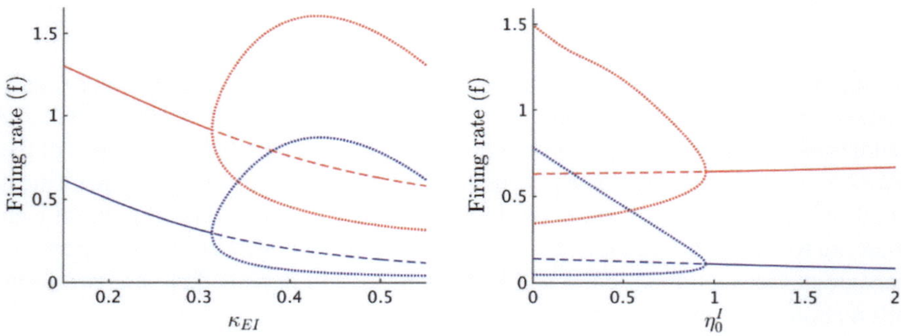

Fig. 4 Corresponding bifurcation diagrams to Fig. 3 under variation in κ_{EI} (left) and η_0^I (right), for both the excitatory (blue) and inhibitory (red) populations. Note that PING rhythms can be terminated by either decreasing the strength of coupling to the excitatory population from the inhibitory population or increasing the natural frequency of the inhibitory population. Parameters as in Fig. 3 with $\kappa_{IE} = 0.9$ for both

The inclusion of self coupling leads to a wide variety of bifurcations, as seen in Fig. 5. As η_0^I is increased we observe the appearance of oscillatory behaviour through a supercritical Hopf bifurcation, which is destroyed at a second supercritical Hopf bifurcation when η_0^I is increased further. Note the appearance/disappearance of period doubling through two period doubling bifurcations on this branch of periodic orbits. We also observe the appearance and disappearance of an isola of periodic orbits through two saddle node bifurcations of periodic orbits. The first saddle node occurs before the second Hopf bifurcation, i.e. there exists two stable periodic orbits

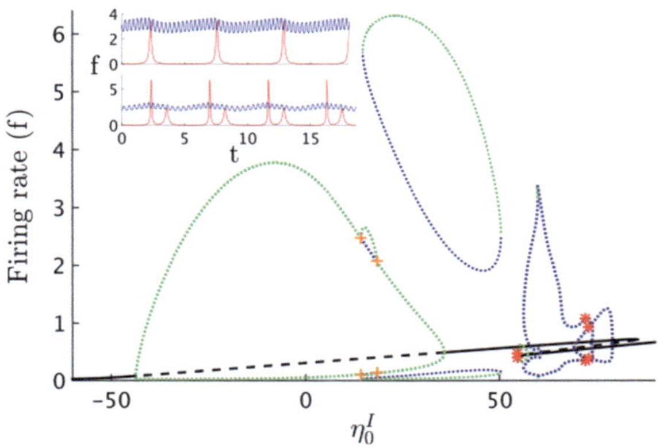

Fig. 5 Bifurcation diagram for η_0^I showing the exotic behaviour of the two population model when self coupling is reintroduced. Solid lines: stable; dashed lines: unstable; green (blue) dotted line: stable (unstable) oscillations; orange crosses: period doubling bifurcations; red stars: torus bifurcations. Of particular interest is the appearance/disappearance of an isola at $\eta_0^I \simeq 15 - 50$. The insets shows the firing rate for the excitatory (blue) and inhibitory (red) populations. In the upper inset $\eta_0^I = -20$ and in the lower inset $\eta_0^I = 25$. Other parameters: $\alpha_{EE} = 1$, $\alpha_{EI} = 0.7, \alpha_{IE} = 1.4$, $\alpha_{II} = 0.4$, $\kappa_{EE} = 1.5$, $\kappa_{EI} = 2$, $\kappa_{IE} = 1$, $\kappa_{II} = 3$, $v_{\text{syn}}^{EI} = 10$, $v_{\text{syn}}^{IE} = -v_{\text{syn}}^{EI} = 8$, $v_{\text{syn}}^{II} = -12$, $\eta_0^E = 20$, $\Delta^E = \Delta^I = 0.5$

for this window of parameter space. Further increasing η_0^I leads to another saddle node bifurcation of periodic orbits, shortly followed by a torus bifurcation and then a saddle-node on invariant circle bifurcation which destroys the unstable branch of the periodic orbit. The stable branch of the periodic orbit is destroyed at a supercritical Hopf bifurcation, as η_0^I is increased further. Along the unstable fixed point branch there are four Hopf bifurcations all of which either create or destroy unstable periodic behaviour. Between the second and third of these bifurcations there are two torus bifurcations, one on each periodic orbit.

The inset in Fig. 5 shows the behaviour of the system for $\eta_0^I = -20$ and $\eta_0^I = 25$ respectively. In both cases the excitatory population has two frequencies and the lower frequency is synchronised to the inhibitory population. For $\eta_0^I = 25$, the system follows the orbit created by the isola, and there are two peaks in the (inhibitory) firing rate per period.

4.2 Event Related Synchronisation and Desynchronisation

Here we show that a time varying input to a single population model given by (26) can both disrupt and enhance the degree of synchrony within the population. We include such a drive under the replacement $\eta_0 \to \eta_0 + J(t)$ (describing a homogeneous drive to the microscopic system), where $J(t)$ is a smoothed rectangular pulse of the form $J(t) = \int_{-\infty}^{t} \eta_D(t-s)A(s)\text{d}s$, for $A(t) = \sigma\Theta(t-T)\Theta(T+\tau-t)$ and $\eta_D(t)$ is an α-function of the form (1) under the replacement $\alpha \to \alpha_D$. In Fig. 6

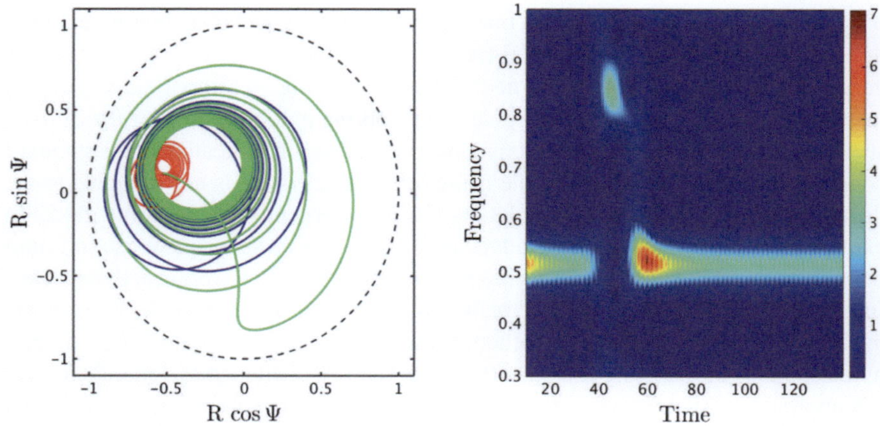

Fig. 6 Left: Phase plane for $Z = Re^{i\Psi}$, demonstrating the behaviour of the system (26) in response to a drive in the form of a smoothed rectangular pulse. The blue curve represents the system before the pulse arrives, as it settles to its non-perturbed dynamics ($t < T$), the red curve demonstrates how the system behaves when the pulse is switched on (at $t = T$) and the green how the system reacts once the drive is switched off (at $t = T + \tau$). Right: The corresponding spectrogram of the synaptic current demonstrating *rebound* (an enhanced spectral peak on cessation of the applied pulse). Parameter values as in Fig. 1, with $T = 40$, $\tau = 12$, $\eta_0 = 21.5$, $\alpha_D = 6$ and $\sigma = 15$

we show that for the inhibitory population considered in Sect. 3.1 operating in its oscillatory regime that such a drive can initially cause a population to desynchronise (during the pulse) though that upon termination of the pulse the system can *rebound* and generate a stronger spectral power peak than seen before the presentation of the pulse (before relaxing back to the undriven periodic orbit). Thus, in contrast to a standard neural mass model, the mean field model (26) is mechanistically able to support the phenomena of ERD and ERS. A more thorough discussion of this observation can be found in Byrne et al. [30], where the model is used to describe magnetoencephalography (MEG) data for movement related beta decrease and post-movement beta rebound [16].

5 Discussion

The desire to understand large scale brain dynamics as observed using EEG, MEG and fMRI has prompted the increasing use of computational models [31]. Many of these approaches, such as exemplified by the Virtual Brain project [14], make use of networks of interconnected neural mass models. However, the inability of a single neural mass model to support the well documented phenomena of ERS and ERD reminds us that these phenomenological models could be improved upon. Of course, building more detailed biophysically realistic models of neurons and their interactions would improve this state of affairs, though at a price. This being not only computational complexity but our ability to interpret the behaviour of very high dimensional models in a meaningful way. The model that we have presented here is very much in the original spirit of neural mass modelling, yet importantly it can be interpreted directly in terms of an underlying spiking model. Moreover the *derived* structure of the macroscopic equations can be viewed as a modification of the standard neural mass framework whereby the firing rate of the system is now coupled to the degree of synchrony within a population. Given the recent success of this model in explaining beta rebound [30], we advocate strongly for its subsequent use in future population-level modelling approaches for understanding *in vivo* brain activity states. Indeed we would like to think that this is a first example of a next generation neural mass model and that there will be others to follow. For phase oscillator single neuron models, this is intrinsically linked to the mathematical challenge of generalising the OA approach, whilst for more general conductance based neurons one might appeal to the McKean-Vlasov-Fokker-Planck approach of Baladron et al. [32]. However, even before extending the model we have presented here to include more biological features (such as action potential generation, dendritic processing, and stochasticity) there is still much to be done in understanding the response of the model to input. Of particular utility would be an understanding of the response to periodic forcing, as this would be a precursor to understanding patterns of phase-locking, clustering, chaos, and the multiplicity of attractors expected at the network level. Moreover, given that neural mass models are themselves the building blocks of neural field models [33] it would be interesting to pursue the analysis of bumps, waves and patterns in continuum versions of the model presented here, along the lines recently developed by Laing for both synaptic and gap junction coupled systems [27, 34].

Acknowledgements SC was supported by the European Commission through the FP7 Marie Curie Initial Training Network 289146, NETT: Neural Engineering Transformative Technologies.

References

1. Wilson, H.R., Cowan, J.D.: Excitatory and inhibitory interactions in localized populations of model neurons. Biophys. J. **12**, 1–24 (1972)
2. Zetterberg, L.H., Kristiansson, L., Mossberg, K.: Performance of a model for a local neuron population. Biol. Cybern. **31**, 15–26 (1978)
3. Lopes da Silva, F.H., Hoeks, A., Smits, H., Zetterberg, L.H.: Model of brain rhythmic activity: the alpha-rhythm of the thalamus. Kybernetik **15**, 27–37 (1974)
4. Lopes da Silva, F.H., van Rotterdam, A., Barts, P., van Heusden, E., Burr, W.: Models of neuronal populations: the basic mechanisms of rhythmicity. Prog. Brain Res. **45**, 281–308 (1976)
5. Jansen, B.H., Rit, V.G.: Electroencephalogram and visual evoked potential generation in a mathematical model of coupled cortical columns. Biol. Cyber. **73**, 357–366 (1995)
6. Wendling, F., Benquet, P., Bartolomei, F., Jirsa, V.: Computational models of epileptiform activity. J. Neurosci. Methods **260**, 233–251 (2016)
7. Dafilis, M.P., Frascoli, F., Cadusch, P.J., Liley, D.T.J.: Chaos and generalised multistability in a mesoscopic model of the electroencephalogram. Phys. D **238**, 1056–1060 (2009)

8. Freeman, W.J.: Tutorial on neurobiology: from single neurons to brain chaos. Int. J. Bifurc. Chaos **2**, 451–482 (1992)
9. Sotero, R.C., Trujillo-Barreto, N.J., Iturria-Medina, Y., Carbonell, F., Jimenez, J.C.: Realistically coupled neural mass models can generate EEG rhythms. Neural Comput. **19**, 478–512 (2007)
10. Spiegler, A., Knösche, T.R., Schwab, K., Haueisen, J., Atay, F.M.: Modeling brain resonance phenomena using a neural mass model. PLoS Comput. Biol. **7**(12), e1002298 (2011)
11. Deco, G., Jirsa, V.K., McIntosh, A.R.: Emerging concepts for the dynamical organization of resting-state activity in the brain. Nature Rev. Neurosci. **12**, 43–56 (2011)
12. Valdes-Sosa, P., Sanchez-Bornot, J.M., Sotero, R.C., Iturria-Medina, Y., Aleman-Gomez, Y., Bosch-Bayard, J., Carbonell, F., Ozaki, T.: Model driven EEG/fMRI fusion of brain oscillations. Hum. Brain Mapp. **30**, 2701–2721 (2009)
13. Moran, R., Pinotsis, D.A., Friston, K.: Neural masses and fields in dynamic causal modeling. Front. Comput. Neurosci. **7**(57), 1–12 (2013)
14. Sanz-Leon, P., Knock, S.A., Spiegler, A., Jirsa, V.K.: Mathematical framework for large-scale brain network modeling in the virtual brain. NeuroImage **111**, 385–430 (2015)
15. Bhattacharya, B.S., Chowdhury, F.N (eds.): Validating Neuro-Computational Models of Neurological and Psychiatric Disorders. Springer, Berlin (2015)
16. Pfurtscheller, G., Lopes da Silva, F.H.: Event-related EEG/MEG synchronization and desynchronization: basic principles. Clin. Neurophys. **110**, 1842–1857 (1999)
17. Ashwin, P., Coombes, S., Nicks, R.: Mathematical frameworks for oscillatory network dynamics in neuroscience. J. Math. Neurosci. **6**(2), (2016)
18. Luke, T.B., Barreto, E., So, P.: Complete classification of the macroscopic behaviour of a heterogeneous network of theta neurons. Neural Comput. **25**, 3207–3234 (2013)
19. Montbrió, E., Pazó, D., Roxin, A.: Macroscopic description for networks of spiking neurons. Phys. Rev. X **5**, 021028 (2015)
20. Spiegler, A., Kiebel, S.J., Atay, F.M., Knösche, T.R.: Bifurcation analysis of neural mass models: impact of extrinsic inputs and dendritic time constants. NeuroImage **52**, 1041–1058 (2010)
21. Touboul, J., Wendling, F., Chauvel, P., Faugeras, O.: Neural mass activity, bifurcations, and epilepsy. Neural Comput. **23**, 3232–3286 (2011)
22. Ermentrout, G.B., Kopell, N.: Parabolic bursting in an excitable system coupled with a slow oscillation. SIAM J. Appl. Math. **46**, 233–253 (1986)
23. Latham, P.E., Richmond, B.J., Nelson, P.G., Nirenberg, S.: Intrinsic dynamics in neuronal networks I theory. J. Neurophysiol. **83**, 808–827 (2000)
24. Pazó, D., Montbrió, E.: Low-dimensional dynamics of populations of pulse-coupled oscillators. Phys. Rev. X **4**, 011009 (2014)
25. Kuramoto, Y.: Collective synchronization of pulse-coupled oscillators and excitable units. Phys. D **50**, 15–30 (1991)
26. Ott, E., Antonsen, T.M.: Low dimensional behavior of large systems of globally coupled oscillators. Chaos **18**, 037113 (2008)
27. Laing, C.R : Phase oscillator network models of brain dynamics. In: Computational Models of Brain and Behavior. Wiley-Blackwell (2016)
28. Ermentrout, G.B.: Simulating, Analyzing, and Animating Dynamical Systems: A Guide to XPPAUT for Researchers and Students. SIAM Books, Philadelphia (2002)
29. Börgers, C., Kopell, N.: Synchronization in networks of excitatory and inhibitory neurons with sparse, random connectivity. Neural Comput. **15**, 509–538 (2003)
30. Byrne, Á., Brookes, M.J., Coombes, S.: A mean field model for movement induced changes in the β rhythm. J. Comput. Neurosci. **43**, 143–158 (2017)
31. Bojak, I., Breakspear, M.: Neuroimaging, neural population models for. In: Encyclopedia of Computational Neuroscience, pp. 1–29. Springer, Berlin (2014)
32. Baladron, J., Fasoli, D., Faugeras, O., Touboul, J.: Mean field description of and propagation of chaos in networks of Hodgkin-Huxley and FitzHugh-Nagumo neurons. J. Math. Neurosci. **2**(10), (2012)

33. Coombes, S., beim Graben, P., Potthast, R., Wright, J.J. (eds). Neural Field Theory. Springer, Berlin (2014)
34. Laing, C.R.: Exact neural fields incorporating gap junctions. SIAM J. Appl. Dyn. Syst. **14**, 1899–1929 (2015)

Unraveling Brain Modularity Through Slow Oscillations

Maurizio Mattia and Maria V. Sanchez-Vives

Abstract The intricate web of connections among the neurons composing the cerebral cortex is the seed of the complexity that our brain is capable to express. Such complexity is organized as it results from a hierarchical and modular organization of the network in which the roles of different cortical areas are distinct. Here, we speculate that such differentiation can be obtained by relying on the granular nature of the cortical surface tiled with 'canonic' modules which in turn can be flexibly tuned to compose diverse mesoscopic networks. The remarkable versatility of these cortical modules is governed by few key parameters like the excitability level and the sensitivity to the accumulated activity-dependent fatigue. These modules are naturally endowed with a rich repertoire of activity regimes which range from quasi-stable dynamics, possibly exploited to store information or provide persistent input to other modules, to collective oscillations reminiscent of the Up/Down activity cycle observed during sleep and deep anesthesia. Finally, we conclude showing that such slow oscillations, spontaneously expressed by the isolated cortex, can provide an ideal experimental framework to infer the dynamical properties of these cortical modules which in turn can inform also on cortical function in other brain states, such as during wakefulness.

The idea of a modular brain in which different competences are shared between differentiated structures and cortical areas is a hallmark of neuroscience developed since its infancy more than one century ago with the pioneering work done by Pierre Paul Broca, Karl Wernicke and Korbinian Brodmann [25]. Brain functions emerge as concerted activity of such ensembles in which a specific role is played by the functional interplay and structural connectivity between cortical areas. Indeed, activity integration and functional segregation can be simultaneously expressed in this

M. Mattia (✉)
Istituto Superiore di Sanità (ISS), Rome, Italy
e-mail: maurizio.mattia@iss.it

M. V. Sanchez-Vives
ICREA, Systems Neuroscience, August Pi i Sunyer Biomedical Research Institute (IDIBAPS), Barcelona, Spain
e-mail: msanche3@clinic.ub.es

F. Corinto and A. Torcini (eds.), *Nonlinear Dynamics in Computational Neuroscience*, PoliTO Springer Series, https://doi.org/10.1007/978-3-319-71048-8_2

network due to its hierarchical organization, giving rise to complex high-dimensional neuronal dynamics [16, 30, 41].

Such organized complexity is not only the result of a well designed connectivity map. Cortical regions have their own specific cytoarchitecture and are associated with different functions [23, 25]. These macroscopic units are known to be composed of rather stereotyped mesoscopic modules – the Mountcastle's minicolumns [38] – tiling the cortical surface and each implementing a 'canonic' circuitry needed to receive input from and send output to other columns and subcortical structures [18, 30]. Hence, the functional and dynamical diversity of cortical regions has a multiscale nature, with a critical role played by some heterogeneity of its mesoscopic components. From the perspective of brain development this is a powerful design paradigm [45], as reusing the same building blocks to compose the cerebral cortex has evident advantages provided that such modules can be suitably tuned to fit the features required by the regions to build. Thus, a still open challenge is to develop models of cortical columns endowed of such flexibility [46] and which at the same time are capable of reproducing the dynamical features observed in in vivo and in vitro experiments.

In what follows, we review the minimal models for such cortical modules capable of reproducing the neurophysiological activity measured at the mesoscopic scale in experiments. We highlight how diverse dynamical modes can emerge that are able to express the heterogeneity of excitability observed across the various cortical regions. We focus on the specific experimental framework of the slow oscillations (SO). SO are an almost periodic alternation at a frequency of less than 1 Hz between high-firing Up periods and almost quiescent Down states. Such alternation is visible both at the single-neuron level, in the collective activity of a local cortical network (Fig. 1), and at the whole brain level. SO are spontaneously displayed by the intact

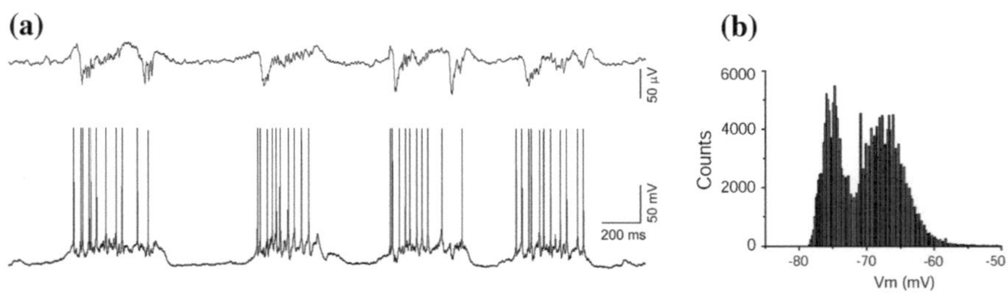

Fig. 1 Brain slow oscillations at microscopic and mesoscopic levels. **a** Simultaneous extracellular (top, intracortical local field potential, LFP) and intracellular (bottom, cellular membrane potental V_m) recordings from the primary auditory cortex of an anesthetized rat. Both signals display the synchronized alternation between a high-firing Up state and a low-activity (almost quiescient) Down state. Rapid fluctuations of LFP reflect the activity at the mesoscopic scale of a cell assembly within a radius of 150 µm around the electrode tip. On the other hand, V_m records the microscopic activity of the single intracellularly recorded neurons. **b** Bimodal distribution of V_m sampled from a long time series of intracellular recordings. The low and high peaks correspond to the membrane potential values (counts) during Down and Up states, respectively. Panels adapted from [51]

brain during deep sleep and anesthesia [11, 14, 49, 59] and in patients with brain lesions [56], all of which are cases where the cortex has been disconnected from the rest of the brain and from the environment. This spontaneous activity has been found in the local cortical network of brain slices [53] and in disconnected cortical slabs [61], strengthening the hypothesis that such slow rhythms are of cortical origin. For these and other reasons, SO have been proposed as a default mode of the cortical network spontaneously arising both in vitro and in vivo in the absence of any stimulation [51]. Such default mode emerges in mammals with less developed cortical circuits, suggesting conservation throughout evolution and thus a relevant function in brain circuits. In the search for function, slow oscillations have been implicated in memory consolidation [17, 31] and synaptic plasticity [33]. Due to their role in slow-wave sleep, they may also be involved in processes such as metabolic and synaptic homeostasis, cellular maintenance and consolidation of information [31, 51, 62, 66, 72].

The Up states represent the start of a propagating wave that is initiated in deep layers of the cortex [11, 50, 53], and then propagates across the cortical network both in vivo [33, 49, 60] and in vitro [10, 70]. Columnar interactions between infragranular and supragranular layers are necessary to keep the appropriate propagation speed, which is modulated by inhibition [12, 52]. SO are highly sensitive to the excitability of the cortical network, which modulates, sometimes parametrically, its different parameters (duration of Up and Down states, firing rates or regularity). This is the case for temperature [47], anesthesia [15], potassium levels [55] and electric fields [20]. SO are thus an ideal experimental framework to study the hierarchical and modular organization of the cerebral cortex and its columnar components in the absence of exogenous inputs and outputs.

1 Nonlinear Dynamics in Neuronal Assemblies

The computational properties of neuronal networks, and hence their capability of transforming and storing information, rely on their nonlinear dynamics. In rather general terms, perception and cognition for instance can be modeled as the transition of the network from one state to another across its phase space [44]. In this framework, nonlinear dynamics is needed in order to make available a rich repertoire of trajectories to be followed in response to the same input but dependent on the state occupied by the system [9, 63]. This is a rather familiar concept among computational neuroscientists since the seminal works by Shun-ichi Amari [3] and John J. Hopfield [24], where the collective emergent properties of a neuronal ensemble were specifically presented, elucidating their potential role in information processing. By properly setting the coupling strength between network units/neurons, a phase transition by 'symmetry breaking' may occur such that the 'symmetric' state in which all the neurons emit spikes at similar low-firing rates is no longer the only available attractor state for the network. Indeed, under this condition a plethora of 'asymmetric' attractor states becomes available. Each of these states has a subset of neurons displaying a persistent high-firing rate of spikes, and implements a short-term memory of the external stimulus that pushes the system to the corresponding valley of the energy landscape. Changes in the environment sensed by the neuronal network can alter such energy landscape. In this way, transitions between states may occur, giving rise to the aforementioned trajectories in the high-dimensional phase space of the network. Dimensionality here is determined by the size of the system, i.e. by the number of units/neurons.

This rather general framework depicted for networks of 'analog' units/neurons, i.e. with a time-varying continuous state variable, can be mapped onto more realistic models of the nervous tissue. Indeed, neurons emit spikes and these can be roughly considered sparse point-like events in time. From this perspective, an effective neuronal model is the leaky integrate-and-firing (IF) neuron [27] in which incoming spikes from the dendritic tree contribute to the production of a current $I(t)$ integrated by the membrane potential $V(t)$ as

$$\frac{dV(t)}{dt} = -\frac{V(t)}{\tau} + I(t), \tag{1}$$

where τ is the decay time due to leakage currents. When $V(t)$ of this point-like (i.e. without shape and spatial extension) neuron reaches a threshold θ, a spike is emitted and sent to downstream cells via the synaptic contacts. This event resets $V(t)$ to H (the reset potential) for a refractory period τ_0. Networks composed of such spiking neurons have nonlinear collective dynamics similar to those highlighted above, and have been often presented as an adequate model of a cortical column provided that both excitatory and inhibitory cells are incorporated [4, 5]. To represent specific neuronal selectivity to different stimuli, as in Hopfield networks, excitatory neurons have to be grouped into homogeneous subpools, each with a potentiated recurrent synaptic coupling J_p (Fig. 2a-top). Focusing on a single 'foreground' pool (fg) and considering the rest of the network as a 'background' (bg) where neurons are coupled through weaker synapses with strength $J_{EE} < J_p$, nonlinear dynamics simply emerge by increasing the so-called potentiation level $w_p = J_p/J_{EE}$. Figure 2a shows simulations of sample networks displaying qualitatively different input-output responses of foreground neurons to the same input (a sudden increase of external current during 250 ms). When synaptic potentiation is small (dark curves), the average firing rate $\nu(t)$ following stimulation reverts to baseline activity levels, forgetting past events. On the other hand, for relatively large w_p, network activity persists in a high-firing state (light curves). In this case local synaptic reverberation is strong enough to allow the cell assembly to sustain such spike emission frequency for long times, thus keeping alive a memory trace of the past stimulation.

The emergence of such collective nonlinear dynamics is fully captured by the 'extended' mean-field theory [4, 21, 67]. This theory assumes that all neurons belonging to the same homogeneous subpool of a cortical module receive stochastic synaptic currents with approximately the same infinitesimal mean $\mu(t)$ and variance $\sigma^2(t)$. This approximation holds under the hypotheses that each neuron receives a large number of spikes per time unit, and that currents due to single presynaptic spikes

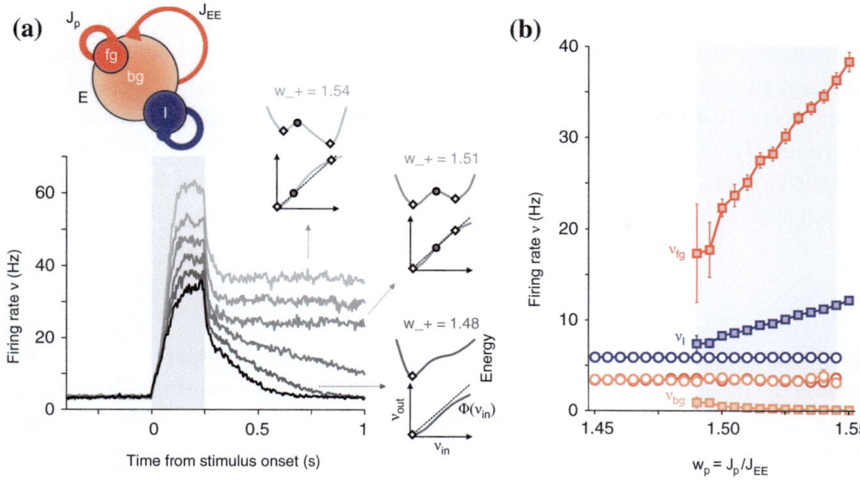

Fig. 2 Rising of point-attractor dynamics in spiking neuron networks by varying the strength of synaptic self-excitation. **a** Firing rate in response to a brief external stimulation (gray bar) of an example of a cortical module. The network is composed of 4000 excitatory and 1000 inhibitory leaky integrate-and-fire (LIF) neurons with random synaptic matrix and 20% connection probability. The excitatory pool is divided into a foreground (fg) and background (bg) subpopulation. Average synaptic strength is stronger between foreground neurons (J_p) than between background cells (J_{EE}). Average firing rate ν_{fg} across 10 stimulations for different potentiation levels $w_p = J_p/J_{EE}$ are coded in gray shades. Insets depict the input-ouput gain function $\Phi(\nu)$ and the effective energy landscape for three different w_p. White diamonds, stable fixed-points (attractors). Filled circles, unstable fixed points. **b** Bifurcation diagram for the network in **a**. For $w_p \geq 1.49$ a region (gray rectangle) where two attractor states at high (dark-filled symbols) and low (white-filled symbols) firing rates coexist. Symbol color as in **a**-top

induce small changes in the postsynaptic membrane potential. All these conditions are well satisfied by cortical neurons and allow modeling the membrane potential as a stochastic diffusive process [64] with invariant statistical properties across neurons [5]. In this theoretical framework, the moments μ and σ^2 are proportional to the rate of spikes $\nu(t)$ generated by the network, and under stationary conditions such frequency turns out to be the inverse of the mean first-passage time (FPT) of the emission threshold θ by $V(t)$. This period is usually named the mean inter-spike interval (ISI), and for the leaky IF neuron in Eq. (1), the mean FPT is the same as the Ornstein-Uhlenbeck process [48, 64]. By knowing the moments of $I(t)$, the stationary output firing rate is given by the following input-output gain function:

$$\Phi(\nu) = \Phi[\mu(\nu), \sigma(\nu)] = \left[\tau_0 + \tau\sqrt{\pi} \int_{\frac{H-\mu\tau}{\sigma\sqrt{\tau}}}^{\frac{\theta-\mu\tau}{\sigma\sqrt{\tau}}} [1 + \mathrm{erf}(x)]\, e^{x^2} dx \right]^{-1}. \quad (2)$$

The gain function $\Phi(\nu)$ is a sigmoid (right insets of Fig. 2a) similar to the one characterizing the units/neurons of the Hopfield network. Fixed-points of this neuronal assembly model are those in which input and output firing rates are self-consistently reproduced: $\nu^* = \Phi(\nu^*)$. Geometrically these are the intersections between the sigmoid and the ν straight line (diamonds and circle in Fig. 2a-right). Depending on w_p there can be from one to three fixed-points, providing the possibility of having multistable dynamics. Local stability can be worked out, and provided that inhibitory synaptic strength is weak enough, fixed-points are attractor states if $\Phi'(\nu^*) < 1$ [34, 35]. An approximated firing rate dynamics capable of reproducing all these features is

$$\tau_\nu \frac{d\nu(t)}{dt} = \Phi[\nu(t)] - \nu(t)\,, \tag{3}$$

where τ_ν is a relaxation constant, to be considered activity-dependent if one wanted to increase the theoretical accuracy [22, 40]. In this framework the firing rate dynamics can be seen as a ball rolling into an energy landscape $E(\nu) = -\int(\Phi(\nu) - \nu)d\nu$, where valleys correspond to stable attractor states (Fig. 2a-right top gray curves). In the example shown in Fig. 2, the mentioned 'symmetry breaking' occurs for $w_p = 1.49$ such that two valleys are simultaneously available beyond this value (Fig. 2b), provided that $w_p < 1.55$. Indeed, above this value, low-firing Down state destabilizes and only one high-firing Up state for the foreground pool (filled red squares) is available. Under a bistable regime like this, such spiking neuron networks have been proposed to be an effective model of the neuronal substrate of cognitive functions such as decision making, short-term working memory and motor planning [4, 36, 67, 68].

2 Widening the Dynamic Repertoire with Fatigue

To have an in silico cortical module closer to the biological reality, an additional ingredient must be included. Indeed, since the seminal work by Edgar D. Adrian [2] during the first half of the past century, it is known that spike frequency declines with time when constant sensory stimulation is delivered. This spike frequency adaptation (SFA) is particularly apparent in excitatory neurons and can take place through several mechanisms. One of these mechanisms can be of synaptic origin (short-term depression) due to a decrease in the probability of neurotransmitters being released when a high presynaptic firing rate persists for periods longer than hundreds of milliseconds [1]. Such 'fatigue' of the synaptic transmission determines a reduction of the currents received by the post-synaptic neurons, thus down-regulating the excitability of the cell assembly. Another fatigue mechanism involves the influx of potassium (K^+) ions whose channels are activated by the accumulation of intracellular calcium (Ca^{2+}) and sodium (Na^+). Since the intracellular concentration of such ions is activity-dependent, the 'hyperpolarizing' K^+ current increases as neurons emit spikes at high frequencies [8, 54]. Due to the long time scales of the ionic gating, such inhibitory currents can last from hundreds of milliseconds to seconds when neurons decrease or stop their firing, as it happens in SO [12, 69]. This brings another distinctive feature of in vivo and in vitro recordings, which is the 'afterhyperpolarization' (AHP) of the membrane potential following the end of the high-firing state (see the hole in the membrane potential trace at the Down state beginnings in Fig. 1a) [52, 54]. These two aforementioned mechanisms are not mutually exclusive, and likely a combination of short-term depression and potassium currents contribute to SFA [42, 43].

Focusing on the activity-dependent K^+ currents as a fatigue process, the single-neuron model can be extended by considering an additional inhibitory current, proportional to the Ca^{2+} or Na^+ ionic concentration $C(t)$, to be integrated in Eq. 1

$$\begin{cases} \frac{dV(t)}{dt} = -\frac{V(t)}{\tau} + I(t) - g\,C(t) \\ \frac{dC(t)}{dt} = -\frac{C(t)}{\tau_c} + \sum_k \delta(t - t_k) \end{cases}, \tag{4}$$

where g is the K^+ channel conductance, τ_c is the relaxation time of the ionic concentration $C(t)$ in absence of spiking activity, and the sum is over the spikes (Dirac's delta) emitted by the neuron at times t_k [22, 26]. To understand the effect of this extension, Fig. 3a shows the measured $\nu(t)$ for the foreground subpools in three simulated networks with different conductances $g > 0$. Networks are set to have a large w_p, such that the only Up state would be accessible if $g = 0$. In an optimal range of g (see squares in the bifurcation diagram of Fig. 3b), Up states spontaneously occur with no need to stimulate the cell assembly as in Fig. 2. Aligning the Up state onsets in all the simulations, it is also apparent that, by increasing g, the maximum firing rate decreases (from light to black). Only for the small enough $g = 0.4$ does high-firing state persist for more than one second; in the other cases the Up state destabilizes within a few hundreds of milliseconds or is never elicited. What is also apparent in these simulations is the SFA phenomenon such that, after the network activation, spike frequency exponentially declines with time.

Mean-field theory is a useful tool also for this improved cortical assembly model. Indeed, provided that $C(t)$ is not too fast compared to $\nu(t)$ dynamics, and its variability across neurons is not too large, an effective dimensional reduction can be worked out [22]. The resulting mean-field dynamics of the network is

$$\begin{cases} \tau_\nu \frac{d\nu}{dt} = \Phi(\nu, c) - \nu \\ \frac{dc}{dt} = -\frac{c}{\tau_c} + \nu \end{cases} \tag{5}$$

where the global 'fatigue' level $c(t)$ is the intracellular concentration $C(t)$ averaged across neurons, and the single-neuron gain function $\Phi(\nu, c)$ now depends both on $\nu(t)$ and $c(t)$. This dynamics can effectively describe also the time evolution of a network in which more than one neuronal population is incorporated, as the case shown in Fig. 3a. Indeed, by resorting to a kind of adiabatic approximation, the

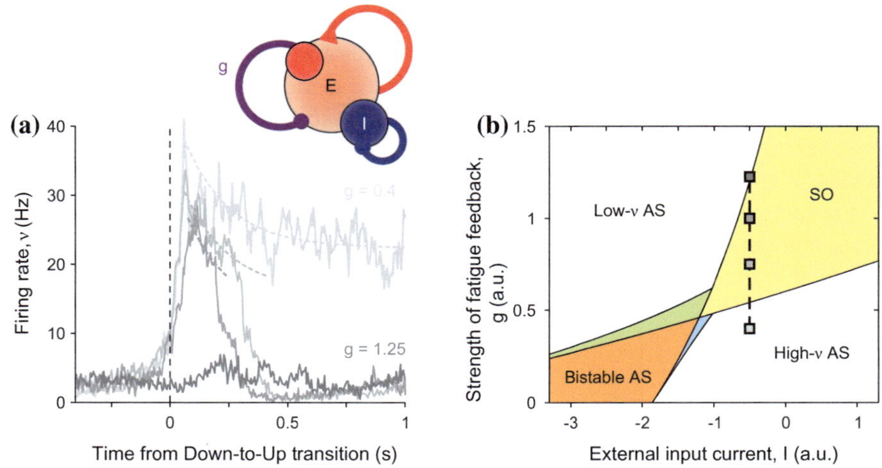

Fig. 3 Spontaneous transition between Up and Down states by introducing activity-dependent self-inhibition. **a** Firing rate $\nu(t)$ of the foreground population varying the strength g of the activity-dependent fatigue feedback (additional purple link in the network sketch). Reference time $t = 0$ s is set to the detected onset of a high-firing Up state, when present. Gray levels, networks with different $g > 0$. Simulated networks as in Fig. 2. Dashed curves, fitted exponential decay of the network activity $\nu(t)$. **b** Bifurcation diagram in which the excitability of the network is modulated by changing an additional external current I, varying also the fatigue feedback g. Different collective activity regimes can be found ranging from single to bistable point-attractors (AS, asynchronous states, white and orange regions) including also stable limit cycles (SO, slow oscillations, yellow region). Green and blue regions are where SO coexist with low- and high-ν AS, respectively. Refer to [22, 29, 37] for additional details. Gray shaded squares, same networks as in **a**. Panels adapted from [37]

other excitatory and inhibitory subpopulations can be seen as rapidly adapting to the dynamics $\nu(t)$ of the excitatory pool that displays the Up/Down oscillations. As a result, the contribution of the other neuronal pools can be integrated into an effective gain function Φ [32]. This chain of simplifications has been proven to be effective in describing quantitatively the dynamical features of the neuronal networks presented here [37].

The interaction between excitation and inhibition like the one in Eq. 5 is known to give rise to an interesting repertoire of dynamical regimes. Indeed, Eq. 5 is equivalent to the dynamics describing the FitzHugh-Nagumo neuron model which is known to display both excitable dynamics and periodic bursting [19, 39]. Not only this, but the same equation is also qualitatively similar to the one introduced in the seminal work by Hugh R. Wilson and Jack D. Cowan [71] that studied the interplay between excitatory and inhibitory neuronal pools. Thus it is not surprising to find the same dynamical richness also in this model of a cortical assembly, as it is shown in Fig. 3b by a section of its bifurcation diagram.

3 The SO Due to a Cortical Relaxation Oscillator

In general, spiking neuron networks are dissipative non-Hamiltonian systems. Thus, the energy landscape metaphor used in Fig. 2 is an approximate representation valid only under specific constraints. Fast inhibition due to inhibitory neurons in the module changes in a few-millisecond time scale and can be assumed to rapidly adapt, balancing the activity of the excitatory pools [58, 65]. This justifies the adiabatic approximation for which the only relevant state variable is the firing rate $\nu(t)$ of the active excitatory pool, i.e. the 'foreground' neurons in Fig. 2. For a slow enough excitatory dynamics, the wandering across an effective energy landscape well represents the network behavior provided that a dissipative component is taken into account. Indeed, friction is needed to dampen the system motion down to reach the bottom of energy valleys.

On the other hand, the slow inhibition underlying the adaptation mechanism introduced above varies across time scales ranging from several hundreds of milliseconds to seconds [8]. Here, the dynamics of excitatory pools are much faster and the K^+ currents can be seen as a quasi-static external inhibition, which slowly shapes the energy landscape, thus keeping such metaphor still as a valid tool to interpret the network behavior from Eq. 5.

In this theoretical framework, SO arise as an alternation between two metastable conditions available only when synaptic recurrent excitation (w_p) is strong enough, as shown in Fig. 2. This is why the region of the bifurcation diagram in Fig. 3b where SO occur can be found only for relatively large g (strength of the inhibitory fatigue-driven feedback) and enough network excitability, which in this case is represented by the intensity of an additional external input current I [12, 22, 28, 37]. The mentioned alternation in simulations can be visualized by plotting simultaneously the firing rate $\nu(t)$ and the fatigue level $c(t)$, as in Fig. 4a. Up state onset always occurs when a minimum c is reached, such that the network is maximally excitable. The transition into the high-firing state gives rise to a rapid accumulation of fatigue which, once it reaches a high threshold level, delivers a strong inhibitory current $-g\,c$. This destabilizes the Up state and elicits a transition towards an almost quiescent Down state during which fatigued excitatory neurons recover as $c(t)$ decays eventually giving rise to the next Up/Down cycle. In the phase plane (c, ν) this shift in phase between firing activity and fatigue level brings the system to have close loop trajectories (orbits) shaped by the nullclines $\dot{\nu} = 0$ and $\dot{c} = 0$ from Eq. 5 (Fig. 4b). Indeed, such orbits follow the stable branches of the ν-nullcline (solid red curves) until the fatigue c 'relaxes' to the threshold levels where high- or low-firing states lose their stability. These relaxation phases are mainly determined by the slowest time scale of the network, τ_c, which in turn makes Up/Down oscillations rather regular.

This dynamical regime reminiscent of a 'relaxation oscillator' in which slow and fast dynamics alternate has been found to match the features of SO measured both in vivo and in vitro [6, 12, 13, 37]. An example is shown in Fig. 4c, where the multi-unit activity (MUA, i.e. the collective firing rate of a local cell assembly) from an extracellular recording in a spontaneously oscillating cortical slice clearly displays a

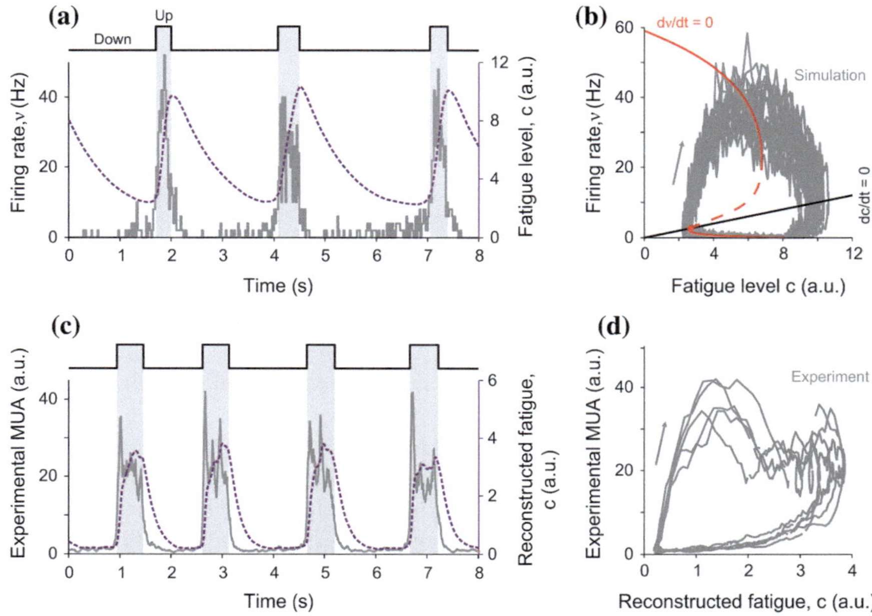

Fig. 4 Nonlinear dynamics of cortical modules underpinning in silico and in vitro slow oscillations. **a** Firing activity $\nu(t)$ of foreground neurons (dark gray line) and their average fatigue level $c(t)$ (purple dashed line) from a simulated network as in Fig. 3 spontaneously displaying an alternation between Down and Up states (white and light gray periods, respectively). **b** Phase plane (c, ν) with the noisy trajectories followed by the simulated network in **a**. Superimposed are the nullclines $\dot{\nu} = 0$ (red) and $\dot{c} = 0$ of the network dynamics under mean-field approximation. Solid and dashed lines, stable and unstable branches of the nullclines, respectively. Red dot, marginally stable fixed-point corresponding to a metastable Down state. **c** In vitro SO. Dark gray line, multi-unit activity (MUA) from an extracellular recording in a ferret cortical slice spontaneously displaying SO [52, 53]. Purple dashed line, reconstructed fatigue level $c(t)$ taking ν as the experimental MUA in Eq. (5) and $\tau_c = 180$ ms, which minimizes trajectory intersections in the phase plane **d** as in previous studies [29]. **d** Phase plane (c, ν) with $\nu =$ MUA and the reconstructed c shown in **c**. Gray lines, reconstructed experimental trajectories. Panel **c** adapted from [37]

rather regular alternation between Up and Down periods, similar to those reproduced in simulated networks (Fig. 4a). Although the experimental value of the 'fatigue' parameter is unknown, identification approaches can be used to uncover τ_c and the other features of Eq. 5 [13, 29]. This, together with the assumption that MUA $= \nu(t)$, allows reconstructing $c(t)$ (dashed purple curve in Fig. 4c). As in simulation, $c(t)$ displays alternating accumulation and relaxation phases locked in time with the Up and Down state onsets. The parameter identification from experiments relies on the expected anti-correlation between ν and c at the beginning of each state, which has been found to be maximal for the real τ_c [29]. As a result, reconstructed trajectories in the phase plane (c, MUA) follow a reproducible orbit and have few intersections, further supporting the hypothesis of a low-dimensionality of the network dynamics. Finally, although the orbits identified from in vitro experiments display some qualitative differences with those from in silico simulations, these findings prove that theoretical models of experimental data can at least in principle be viable to provide quantitative hints about the dynamical underpinnings of SO.

4 Conclusions

In this Chapter, we focused on the emergent properties of the collective dynamics in a cortical module. This unit is the basic element of a wider multi-modular neuronal network in which other brain structures and cortical areas are involved. By dealing with SO, this aspect is obviously relevant, as they are the local expression of the 'slow-wave activity', a multiscale phenomenon whose origin is not yet fully established. Indeed, it is still debated whether slow-wave activity initiates from the cortex or the thalamus [14, 57, 60], although an increasing amount of evidence supports its cortical origin [53, 60, 61].

We summarized the remarkable versatility of the aforementioned mesoscopic neuronal module. With few changes of its parameters, the network can express quasi-stable dynamics to store information or provide persistent input to other modules, and at the same time can behave like an autonomous oscillator. All of this emerges as a collective phenomenon, provided that the system is positioned close to a sweet spot in the bifurcation diagram of Fig. 3b. Around it, such dynamical regimes can be tuned by small changes of key parameters like the excitability level – modulated by the strength of the excitatory synaptic reverberation or by an additional input current – and the sensitivity to the accumulated activity-dependent fatigue.

The dynamical richness of these simple and stereotyped building blocks can be further widened when they are connected together to compose macroscopic cortical areas. The cerebral cortex can thus been seen as a network of such modules and, depending on their working regime, different collective properties can be produced, giving rise to the capability of transforming sensory inputs, but also of spontaneously expressing traveling waves. Thus, understanding SO in such a multi-modular cortical web will provide hints on the dynamical properties that the same network displays in other brain states, such as during wakefulness. Furthermore, characterizing the heterogeneity of the relaxation oscillation features observed under slow-wave activity from different brain areas can also help to understand the root of the functional diversity of cortical areas [7, 49].

Acknowledgements We thank C. Gonzales-Liencres for comments on this manuscript. This work was in part supported by CORTICONIC (EC FP7 grant 600806) and by the EU Horizon 2020 Research and Innovation Program under HBP SGA1 (grant 720270) to MM and MVSV, by the Spanish Ministerio de Ciencia e Innovación (BFU2017-85048-R) and by PCIN-2015-162-C02-01 (FLAG ERA) to MVSV.

References

1. Abbott, L.F., Regehr, W.G.: Synaptic computation. Nature **431**(7010), 796–803 (2004)
2. Adrian, E.D.: The impulses produced by sensory nerve endings: Part I. J. Physiol. **61**(1), 49–72 (1926)
3. Amari, S.I.: Learning patterns and pattern sequences by self-organizing nets of threshold elements. IEEE Trans. Comput. **C–21**(11), 1197–1206 (1972). https://doi.org/10.1109/T-C.1972.223477
4. Amit, D.J., Brunel, N.: Model of global spontaneous activity and local structured activity during delay periods in the cerebral cortex. Cereb. Cortex **7**(3), 237–252 (1997)
5. Amit, D.J., Tsodyks, M.: Quantitative study of attractor neural network retrieving at low spike rates: I. substratespikes, rates and neuronal gain. Netw. Comput. Neural **2**(3), 259–273 (1991). https://doi.org/10.1088/0954-898X/2/3/003
6. Bazhenov, M., Timofeev, I., Steriade, M., Sejnowski, T.J.: Model of thalamocortical slow-wave sleep oscillations and transitions to activated States. J. Neurosci. **22**(19), 8691–8704 (2002)
7. Bettinardi, R.G., Tort-colet, N., Ruiz-mejias, M., Sanchez-vives, M.V., Deco, G.: Gradual emergence of spontaneous correlated brain activity during fading of general anesthesia in rats: evidences from fMRI and local field potentials. NeuroImage **114**, 185–198 (2015). https://doi.org/10.1016/j.neuroimage.2015.03.037
8. Bhattacharjee, A., Kaczmarek, L.K.: For K+ channels, Na+ is the new Ca2+. Trends Neurosci. **28**(8), 422–428 (2005). https://doi.org/10.1016/j.tins.2005.06.003
9. Buonomano, D.V., Maass, W.: State-dependent computations: spatiotemporal processing in cortical networks. Nat. Rev. Neurosci. **10**(2), 113–125 (2009). https://doi.org/10.1038/nrn2558
10. Capone, C., Rebollo, B., Muñoz, A., Illa, X., Del Giudice, P., Sanchez-Vives, M. V., Mattia, M.: Slow waves in cortical slices: how spontaneous activity is shaped by laminar structure. Cereb. Cortex (November), 1–17 (2017). https://doi.org/10.1093/cercor/bhx326
11. Chauvette, S., Volgushev, M., Timofeev, I.: Origin of active states in local neocortical networks during slow sleep oscillation. Cereb. Cortex **20**(11), 2660–2674 (2010). https://doi.org/10.1093/cercor/bhq009
12. Compte, A., Sanchez-Vives, M.V., McCormick, D.A., Wang, X.J.: Cellular and network mechanisms of slow oscillatory activity (<1 Hz) and wave propagations in a cortical network model. J. Neurophysiol. **89**(5), 2707–2725 (2003). https://doi.org/10.1152/jn.00845.2002
13. Curto, C., Sakata, S., Marguet, S., Itskov, V., Harris, K.D.: A simple model of cortical dynamics explains variability and state dependence of sensory responses in urethane-anesthetized auditory cortex. J. Neurosci. **29**(34), 10,600–10612 (2009). https://doi.org/10.1523/JNEUROSCI.2053-09.2009
14. David, F., Schmiedt, J., Taylor, H., Orban, G., Di Giovanni, G., Uebele, V., Renger, J., Lambert, R., Leresche, N., Crunelli, V.: Essential thalamic contribution to slow waves of natural sleep. J. Neurosci. **33**(50), 19599–19610 (2013). https://doi.org/10.1523/JNEUROSCI.3169-13.2013
15. Deco, G., Martí, D., Ledberg, A., Reig, R., Sanchez-Vives, M.V.: Effective reduced diffusion-models: a data driven approach to the analysis of neuronal dynamics. PLoS Comput. Biol. **5**(12), e1000,587 (2009). https://doi.org/10.1371/journal.pcbi.1000587
16. Deco, G., Tononi, G., Boly, M., Kringelbach, M.L.: Rethinking segregation and integration: contributions of whole-brain modelling. Nat. Rev. Neurosci. **16**(7), 430–439 (2015). https://doi.org/10.1038/nrn3963
17. Diekelmann, S., Born, J.: The memory function of sleep. Nat. Rev. Neurosci. **11**(2), 114–126 (2010). https://doi.org/10.1038/nrn2762
18. Douglas, R.J., Martin, K.A.C.: Neuronal circuits of the neocortex. Annu. Rev. Neurosci. **27**, 419–451 (2004). https://doi.org/10.1146/annurev.neuro.27.070203.144152
19. FitzHugh, R.: Impulses and physiologcal states in theoretical models of nerve membrane. Biophys. J. **1**, 445–466 (1961). https://doi.org/10.1016/S0006-3495(61)86902-6
20. Fröhlich, F.: Endogenous electric fields may guide neocortical network activity. Neuron **67**(1), 129–143 (2010). https://doi.org/10.1016/j.neuron.2010.06.005

21. Fusi, S., Mattia, M.: Collective behavior of networks with linear (VLSI) integrate-and-fire neurons. Neural Comput. **11**(3), 633–652 (1999)
22. Gigante, G., Mattia, M., Del Giudice, P.: Diverse population-bursting modes of adapting spiking neurons. Phys. Rev. Lett. **98**(14), 148,101 (2007). https://doi.org/10.1103/PhysRevLett.98.148101
23. Glasser, M.F., Coalson, T.S., Robinson, E.C., Hacker, C.D., Harwell, J., Yacoub, E., Ugurbil, K., Andersson, J., Beckmann, C.F., Jenkinson, M., Smith, S.M., Van Essen, D.C.: A multi-modal parcellation of human cerebral cortex. Nature **536**(7615), 171–178 (2016). https://doi.org/10.1038/nature18933
24. Hopfield, J.J.: Neural networks and physical systems with emergent collective computational abilities. Proc. Natl. Acad. Sci. USA **79**(8), 2554–2558 (1982). https://doi.org/10.1073/pnas.79.8.2554
25. Kandel, E.R., Schwartz, J.H., Jessell, T.M., Siegelbaum, S.A., Hudspeth, A.: Principles of Neural Science, vol. 4. McGraw-Hill, New York (2000)
26. Koch, C.: Biophysics of Computation: Information Processing in Single Neurons. Oxford University Press, New York (1999)
27. Lapicque, L.: Recherches quantitatives sur lexcitation électrique des nerfs traitée comme une polarisation. J. Physiol. Pathol. Générale **9**, 567–578; **9**, 620–635 (1907)
28. Latham, P.E., Richmond, B.J., Nelson, P.G., Nirenberg, S.: Intrinsic dynamics in neuronal networks. I. Theory. J. Neurophysiol. **83**(2), 808–827 (2000)
29. Linaro, D., Storace, M., Mattia, M.: Inferring network dynamics and neuron properties from population recordings. Front. Comput. Neurosci. **5**, 1–17 (2011). https://doi.org/10.3389/fncom.2011.00043
30. Markov, N.T., Ercsey-Ravasz, M., Van Essen, D.C., Knoblauch, K., Toroczkai, Z., Kennedy, H.: Cortical high-density counterstream architectures. Science **342**(6158), 1238,406 (2013). DOI https://doi.org/10.1126/science.1238406
31. Marshall, L., Helgadóttir, H., Mölle, M., Born, J.: Boosting slow oscillations during sleep potentiates memory. Nature **444**(7119), 610–613 (2006). https://doi.org/10.1038/nature05278
32. Mascaro, M., Amit, D.J.: Effective neural response function for collective population states. Netw. Comput. Neural **10**(4), 351–373 (1999)
33. Massimini, M., Huber, R., Ferrarelli, F., Hill, S.L., Tononi, G.: The sleep slow oscillation as a traveling wave. J. Neurosci. **24**(31), 6862–6870 (2004). https://doi.org/10.1523/JNEUROSCI.1318-04.2004
34. Mattia, M., Del Giudice, P.: Population dynamics of interacting spiking neurons. Phys. Rev. E **66**(5 Pt 1), 051,917 (2002)
35. Mattia, M., Del Giudice, P.: Finite-size dynamics of inhibitory and excitatory interacting spiking neurons. Phys. Rev. E **70**, 052,903 (2004). DOI https://doi.org/10.1103/PhysRevE.70.052903
36. Mattia, M., Pani, P., Mirabella, G., Costa, S., Del Giudice, P., Ferraina, S.: Heterogeneous attractor cell assemblies for motor planning in premotor cortex. J. Neurosci. **33**(27), 11155–11168 (2013). https://doi.org/10.1523/JNEUROSCI.4664-12.2013
37. Mattia, M., Sanchez-Vives, M.V.: Exploring the spectrum of dynamical regimes and timescales in spontaneous cortical activity. Cogn. Neurodyn. **6**(3), 239–250 (2012). https://doi.org/10.1007/s11571-011-9179-4
38. Mountcastle, V.B.: The columnar organization of the neocortex. Brain **120**(Pt 4), 701–722 (1997)
39. Nagumo, J., Arimoto, S., Yoshizawa, S.: An active pulse transmission line simulating nerve axon. Proc. IRE **50**, 2061–2070 (1962). https://doi.org/10.1109/JRPROC.1962.288235
40. Ostojic, S., Brunel, N.: From spiking neuron models to linear-nonlinear models. PLoS Comput. Biol. **7**(1), e1001,056 (2011). https://doi.org/10.1371/journal.pcbi.1001056
41. Park, H.J., Friston, K.J.: Structural and functional brain networks: from connections to cognition. Science **342**(6158), 1238,411 (2013). https://doi.org/10.1126/science.1238411
42. Puccini, G.D., Sanchez-Vives, M.V., Compte, A.: Selective detection of abrupt input changes by integration of spike-frequency adaptation and synaptic depression in a computational network model. J. Physiol. Paris **100**(1), 1–15 (2006)

43. Puccini, G.D., Sanchez-Vives, M.V., Compte, A.: Integrated mechanisms of anticipation and rate-of-change computations in cortical circuits. PLoS Comput. Biol. **3**(5), e82 (2007)
44. Rabinovich, M.I., Huerta, R., Laurent, G.: Transient dynamics for neural processing. Science **321**(5885), 48–50 (2008). https://doi.org/10.1126/science.1155564
45. Rakic, P.: Specification of cerebral cortical areas. Science **241**(4862), 170–176 (1988)
46. Rakic, P.: Confusing cortical columns. Proc. Natl. Acad. Sci. USA **105**(34), 12099–12100 (2008). https://doi.org/10.1073/pnas.0807271105
47. Reig, R., Mattia, M., Compte, A., Belmonte, C., Sanchez-Vives, M.V.: Temperature modulation of slow and fast cortical rhythms. J. Neurophysiol. **103**(3), 1253–1261 (2010). https://doi.org/10.1152/jn.00890.2009
48. Ricciardi, L.M., Sacerdote, L.: The Ornstein-Uhlenbeck process as a model for neuronal activity. I. mean and variance of the firing time. Biol. Cybern. **35**(1), 1–9 (1979)
49. Ruiz-Mejias, M., Ciria-Suarez, L., Mattia, M., Sanchez-Vives, M.V.: Slow and fast rhythms generated in the cerebral cortex of the anesthetized mouse. J. Neurophysiol. **106**(6), 2910–2921 (2011). https://doi.org/10.1152/jn.00440.2011
50. Sakata, S., Harris, K.D.: Laminar structure of spontaneous and sensory-evoked population activity in auditory cortex. Neuron **64**(3), 404–418 (2009). https://doi.org/10.1016/j.neuron.2009.09.020
51. Sanchez-Vives, M.V., Mattia, M.: Slow wave activity as the default mode of the cerebral cortex. Arch. Ital. Biol. **152**(2-3), 147–155 (2014). DOI https://doi.org/10.12871/000298292014239
52. Sanchez-Vives, M.V., Mattia, M., Compte, A., Perez-Zabalza, M., Winograd, M., Descalzo, V.F., Reig, R.: Inhibitory modulation of cortical up states. J. Neurophysiol. **104**(3), 1314–1324 (2010). https://doi.org/10.1152/jn.00178.2010
53. Sanchez-Vives, M.V., McCormick, D.A.: Cellular and network mechanisms of rhythmic recurrent activity in neocortex. Nat. Neurosci. **3**(10), 1027–1034 (2000). https://doi.org/10.1038/79848
54. Sanchez-Vives, M.V., Nowak, L.G., McCormick, D.A.: Cellular mechanisms of long-lasting adaptation in visual cortical neurons in vitro. J. Neurosci. **20**(11), 4286–4299 (2000). 20/11/4286. https://doi.org/10.1523/JNEUROSCI.20-11-04286.2000
55. Sancristóbal, B., Rebollo, B., Boada, P., Sanchez-Vives, M.V., Garcia-Ojalvo, J.: Collective stochastic coherence in recurrent neuronal networks. Nat. Phys. **12**, 1–8 (2016). https://doi.org/10.1038/nphys3739
56. Sarasso, S., Rosanova, M., Casali, A.G., Casarotto, S., Fecchio, M., Boly, M., Gosseries, O., Tononi, G., Laureys, S., Massimini, M.: Quantifying cortical eeg responses to tms in (un) consciousness. Clin. EEG Neurosci. **45**(1), 40–49 (2014)
57. Sheroziya, M., Timofeev, I.: Global intracellular slow-wave dynamics of the thalamocortical system. J. Neurosci. **34**(26), 8875–8893 (2014). https://doi.org/10.1523/JNEUROSCI.4460-13.2014
58. Shu, Y., Hasenstaub, A., McCormick, D.A.: Turning on and off recurrent balanced cortical activity. Nature **423**(6937), 288–293 (2003). https://doi.org/10.1038/nature01616
59. Steriade, M., Nuñez, A., Amzica, F.: A novel slow (< 1 Hz) oscillation of neocortical neurons in vivo: depolarizing and hyperpolarizing components. J. Neurosci. **13**(8), 3252–3265 (1993)
60. Stroh, A., Adelsberger, H., Groh, A., Rühlmann, C., Fischer, S., Schierloh, A., Deisseroth, K., Konnerth, A.: Making waves: initiation and propagation of corticothalamic Ca2+ waves in vivo. Neuron **77**(6), 1136–1150 (2013). https://doi.org/10.1016/j.neuron.2013.01.031
61. Timofeev, I., Grenier, F., Bazhenov, M., Sejnowski, T.J., Steriade, M.: Origin of slow cortical oscillations in deafferented cortical slabs. Cereb. Cortex **10**(12), 1185–1199 (2000)
62. Tononi, G., Cirelli, C.: Sleep and the price of plasticity: from synaptic and cellular homeostasis to memory consolidation and integration. Neuron **81**(1), 12–34 (2014). https://doi.org/10.1016/j.neuron.2013.12.025
63. Tsuda, I.: Toward an interpretation of dynamic neural activity in terms of chaotic dynamical systems. Behav. Brain Sci. **24**(5), 793–810 (2001). (discussion 810–848)
64. Tuckwell, H.C.: Introduction to Theoretical Neurobiology; Nonlinear and Stochastic Theories, vol. 2; vol. 8. Cambridge University Press, Cambridge (1988)

65. Van Vreeswijk, C., Sompolinsky, H.: Chaos in neuronal networks with balanced excitatory and inhibitory activity. Science **274**(5293), 1724–1726 (1996)
66. Vyazovskiy, V.V., Harris, K.D.: Sleep and the single neuron: the role of global slow oscillations in individual cell rest. Nat. Rev. Neurosci. **14**(6), 443–451 (2013). https://doi.org/10.1038/nrn3494
67. Wang, X.J.: Synaptic basis of cortical persistent activity: the importance of NMDA receptors to working memory. J. Neurosci. **19**(21), 9587–9603 (1999)
68. Wang, X.J.: Probabilistic decision making by slow reverberation in cortical circuits. Neuron **36**(5), 955–968 (2002)
69. Wang, X.J., Liu, Y., Sanchez-Vives, M.V., McCormick, D.A.: Adaptation and temporal decorrelation by single neurons in the primary visual cortex. J. Neurophysiol. **89**(6), 3279–3293 (2003)
70. Wester, J.C., Contreras, D.: Columnar interactions determine horizontal propagation of recurrent network activity in neocortex. J. Neurosci. **32**(16), 5454–5471 (2012). https://doi.org/10.1523/JNEUROSCI.5006-11.2012
71. Wilson, H.R., Cowan, J.D.: Excitatory and inhibitory interactions in localized populations of model neurons. Biophys. J. **12**(1), 1–24 (1972). https://doi.org/10.1016/S0006-3495(72)86068-5
72. Xie, L., Kang, H., Xu, Q., Chen, M.J., Liao, Y., Thiyagarajan, M., O'Donnell, J., Christensen, D.J., Nicholson, C., Iliff, J.J., Takano, T., Deane, R., Nedergaard, M.: Sleep drives metabolite clearance from the adult brain. Science **342**(6156), 373–377 (2013). https://doi.org/10.1126/science.1241224

Characterization of Neural Signals in Preclinical Studies of Neural Plasticity Using Nonlinear Time Series Analysis

Fabio Vallone, Matteo Caleo and Angelo Di Garbo

Abstract The capacity of the brain to change its basic functions and structures is called neural plasticity. Neural plasticity is one of the most challenging themes in neuroscience and its comprehension may lead to fundamental understanding on brain dynamics. Here, we have characterized the intra and inter regional neuroplastic changes in animal models (mice) of different diseases (i.e. stroke and epilepsy) and sensorimotor stimulation induced by environmental enrichment conditions through the quantitative linear and nonlinear analysis of electrophysiological signals (Local Field Potentials, LFPs). Various properties characterizing LFPs such as power spectra, scaling behavior and interdependence have been quantified. These characterizations were able to discriminate between different experimental conditions, thus providing a good set of quantities that could be useful as biomarkers in medical diagnostics. In particular, we reported some cases in which nonlinear time series analysis reveals effects that are not detected by linear methods. For the epileptic mice, the spectral analysis has shown that epileptic activity determines a power redistribution among the different neurophysiological bands. Symbolic Transfer Entropy measure indicates a greater driving influence of the focal epileptic side on activity in the contra-lateral hemisphere, while Granger causality measures fails at detecting it. Inter-hemispheric functional coupling within delta band (0.5–4 Hz) was reduced in homotopic Pre-Motor Areas of ischemic animals, as shown by a statistically significant decrease in the mutual information measures (not captured by cross-correlation index). Finally, we estimated the scaling properties of LFPs recorded from freely-moving mice reared in environmental enrichment (EE) and standard condition (SC) by using an integrated approach combining three different techniques: the Higuchi method, DFA and spectral analysis. Our results indicated the existence of a particular

F. Vallone · A. Di Garbo (✉)
Institute of Biophysics-CNR, Pisa, Italy
e-mail: angelo.digarbo@pi.ibf.cnr.it

F. Vallone
The Biorobotics Institute-SSSUP, Pontedera, Italy
e-mail: fabiovallone@gmail.com

M. Caleo
Neuroscience Institute-CNR, Pisa, Italy
e-mail: caleo@in.cnr.it

F. Corinto and A. Torcini (eds.), *Nonlinear Dynamics in Computational Neuroscience*, PoliTO Springer Series, https://doi.org/10.1007/978-3-319-71048-8_3

power spectrum scaling law $1/f^{\beta}$ with $\beta < 0$ for low frequencies ($f < 4$ Hz) for both SC and EE rearing conditions. Notably, the difference between scaling exponents in EE and SC for individual cortical regions (M2) and (V1) was not statistically significant. Altogether, these findings shed light on the mechanism involved in neocortical plasticity suggesting both robust plasticity of transcallosal interactions and intra-hemispheric rearrangement of the local neural activities in normal and pathological brain conditions.

1 Introduction

Understanding how the brain works is one the greatest challenge of our century in which physicists can provide new tools for data analysis as well as general principles of brain function [8, 49].

Experiments on the electrical activity of the brain typically provide signals representing the spatiotemporal neuronal activities of thousands of interacting neurons embedded in a extracellular medium [12]. There are various type of recordings methods and the most employed are: Electroencephalography (EEG), Magnetoencephalography (MEG), Electrocorticography (ECoG), Local Field Potentials (LFP) (see [12] and references therein). Each of the above techniques presents advantages and limitations that we describe briefly in the following.

EEG is one of the most widely used methods and it has the value to be a non-invasive technique suitable for studies on human subjects. Technically, millimeter-scale electrodes are placed on the scalp to record the integrated electrical activity of a brain region of size around $10\,\text{cm}^2$. Unfortunately, the direct relationship between neuronal activity and EEG recordings it is not well defined because the electrical activity, generated by thousands of neurons, must diffuse through various media (cerebrospinal fluid, dura matter, cranium, muscle and skin) that distort and attenuate the signals.

Another non-invasive technique suitable for human subjects is MEG. Experiments using MEG are able to measure outside the skull the tiny magnetic field generated by neuronal currents. MEG has a high spatiotemporal resolution ~ 1 ms, 2–3 mm, however the distortion due to the media is still present as in the EEG case (even if much lower than for EEG signals). To avoid the distortion due to the media, we must directly record from the cerebral cortex. A technique usually employed in epileptic patients is ECoG (see for example [47]). This is an invasive method that use electrodes placed directly on the surface of the cerebral cortex allowing a spatial resolution of $<5\,\text{mm}^2$. However, with ECoG we are not able to record in deeper layers than superficial areas and another type of recordings called LFPs must be employed. Even if LFP recordings are obtained through an invasive procedure, suitable only for animal models or patients undergoing specific brain surgery, it is believed that LFP is the most informative signal to study local cortical electrical activity [12]. For example, using bipolar electrodes one is able to sample the electrical activity generated by localized population of neurons within ~ 1 mm and time resolution of the order of milliseconds. Notably, LFP signals do not suffer from distortion effects thus providing a direct measure of the extracellular field generated by all transmembrane currents. In particular, the main contribution to the LFP recording is due to the synaptic currents (since they are slow events that can be more easily synchronized than fast events), although we know the influence of other factors such as fast (Na^+) action potentials, Calcium spikes, spike afterhyperpolarization, gap junctions, neuron-glia interactions and ephatic effects can contribute to the LFP signal. It is worth to mention that besides the methods that measure directly the electrical activity of the brain, also imaging techniques are widely employed. One of the most used, especially in cognitive neuroscience, is functional Magnetic Resonance Imaging (fMRI) that provides an indirect measure of the electrical activity of a brain area through haemodynamic response (for example blood-oxygen-level dependent measure). The main advantage of fMRI lies in its non-invasive nature and a relatively good spatial resolution, however the strong limitation of the technique is the difficulty to relate it to the underlying neural activity and the poor temporal resolution (see [32] and references therein).

Generally speaking, all the above mentioned recording methods produce time series, i.e. values of certain variables in time, that need to be analyzed in order to understand the underlying neurophysiological processes responsible for their generation. This is a very hard task to accomplish [1, 9, 28] since a lot of factors (e.g. noise in the system, finite length of the data, nonstationarities, sparse observations) can introduce a bias in the considered analysis. However, even if we are not able to univocally determine the dynamics of the system we can apply nonlinear time series analysis technique to other tasks such as the characterization of signals providing through suitable measures a set of number that reflects properties of the time series. The characterization of signals turned out to be very useful in many areas of science and for our interests in medical diagnostic and preclinical studies [2, 14, 33, 58, 59]. Indeed, the quantitative characterization of signals could allow clinicians to discriminate if a patient suffer (or will suffer) from some neural pathology even without showing any specific signs or could help doctors to quantify the severity of the disease, improving in this way our diagnostic and prognosis capabilities. Many researches are aimed at paving the way to this future approach of clinics, in particular exploiting animal models that can provide a solid background of experimental experience to be translated to human cases [2, 59]. For instance, our group has related the spontaneous recovery of forelimb motor function to specific value of inter-hemispheric correlation measures in a mouse model of stroke [59]. In other studies in humans, it was shown that the hand (or feet) movement is accompanied by an increase of the coupling between the respective primary motor and premotor areas of both hemispheres [62]. It is important to point out that nonlinear time series analysis techniques are needed together with linear methods to characterize brain signals. Indeed, it is very hard to think of the brain as a linear system knowing that its principal component, i.e. one of the $\sim 10^{11}$ neurons in the cortex, behaves as a nonlinear oscillator described by Hodgkin Huxley equations [24]. Linear methods are able to detect properties of the time series up to second order statistical quantities, indeed usually these techniques assume that the time series can be modeled by autoregressive models which are completely specified by their mean and variance. On the other hand, nonlinear time series analysis explore higher statistical moments than second order statistics allowing to quantify properties that cannot be characterized by linear method [1, 9, 28]. However, it is also possible that linear methods can perform a better analysis than nonlinear methods (see for example [23]). Thus our point of view is that both linear and nonlinear methods must be combined in the analysis of time series. Here, we provided a synopsis of three studies [58–60] were we focused on rodent models (mice) of epilepsy, stroke and Environmental Enrichment (EE) conditions in which LFPs were recorded in different brain areas (see Sect. 3 for details). LFPs were analyzed by several methods of linear and nonlinear univariate or bivariate time series analysis [1, 9, 18, 28, 46] aiming at finding common dynamical indexes to be used as indicators of local neural activity and functional coupling between different areas. Longitudinal variations in the values of such indicators can be employed to quantify the level of synaptic reorganization and plasticity during rehabilitation processes, medical therapy and for identifying suitable prognostic indicators.

We will describe in Sect. 2 the methods of nonlinear time series analysis that we used in our studies. Next, in Sect. 3 we will present some results from the analysis of LFPs, in particular, we will report some cases in which nonlinear time series analysis reveals effects that are not detected by linear methods. Finally, in Sect. 4 we will give the discussion and conclusion.

2 Linear and Nonlinear Time Series Analysis Methods

Let us consider a dynamical system described by a set of coupled equations $\frac{d\mathbf{x}}{dt} = F(\mathbf{x})$, where $\mathbf{x} \in R^n$ is the state of the system at time t. In most cases we do not know these equation (and this is particularly true when dealing with biological systems) and what we can do is just to collect the measurements of a given quantity and try to use these measures to recover useful information on the studied system. More formally, we can introduce the concept of a measure function M as a mapping from the whole phase space of the system to R: i.e. the function $M(\mathbf{x}(t)) = s(t)$. Then, choosing a suitable sampling interval Δt, we get the time series $s(n\Delta t)$, $n = 1, 2, \ldots, N$ containing implicit information on the state of the system. Thus, one of the main aims of nonlinear time series analysis is to solve this inverse problem trying to recover information on the state of the system and consequently on the basic processes driving it. The most important concept in nonlinear time series analysis useful to recover the missing information due to incomplete observations is the notion of state-space reconstruction. The idea is to use time-delayed copies of the available signal $s(t)$ to reconstruct the full dynamics of the considered nonlinear system. Then, a delay reconstruction with delay time τ and embedding dimension m is defined as

$$\mathbf{R}(t) = [s(t), \ldots, s(t - (m - 2)t), s(t - (m - 1)t)]. \tag{1}$$

It can be demonstrated that the reconstruction is topological identical to the full dynamics (see [1, 9, 28]and references therein), however the applicability of state-space reconstruction in real data can be very problematic since it requires infinite, noise free trajectories of a dynamical system. In practice, one usually wants to characterize a signal by computing characteristic numbers that distinguish between it and some other signals. Thus the estimation of characteristic properties from a large set of data is one of the main goals of both linear and nonlinear time series analysis.

A paradigmatic example where this protocol applies is that of neurophysiological experiments [18, 46] in which we want to understand the properties of a system (e.g. a region of the brain) having access only to some variable of the whole system (for example the potential generated by transmembrane currents in a region of the brain, see for example a LFP signal in Fig. 1). Here, we focused on some applications of linear and nonlinear time series methods that we employed in our studies [58–60], such as spectral analysis, coupling between time series and scaling properties .

2.1 Spectral Methods

It has been observed that the brain produces rhythmic activity with distinct functional meaning [11]. For example, it was observed that low frequencies oscillations ($f < 4$) Hz were enhanced during slow wave sleep compared to wakefulness [57]. Moreover, it is believed that high frequency oscillations $f > 30$ Hz, i.e. "gamma-rhythm", modulate the inter-regional communication between brain areas [13]. To investigate the rhythmic brain activity we can apply the standard Fourier analysis to the neural recordings by calculating the corresponding Fourier amplitudes a_{f_k}. From the a_{f_k} we can calculate the power spectrum of the signals using the Welch method of averaging modified periodograms [1, 28] and the power content of the standard neurophysiological spectral bands: $\delta < 4$ Hz, $\theta = (4-8)$ Hz, $\alpha = (8-12)$ Hz, $\beta = (12-30)$ Hz, $\gamma > 30$ Hz [11].

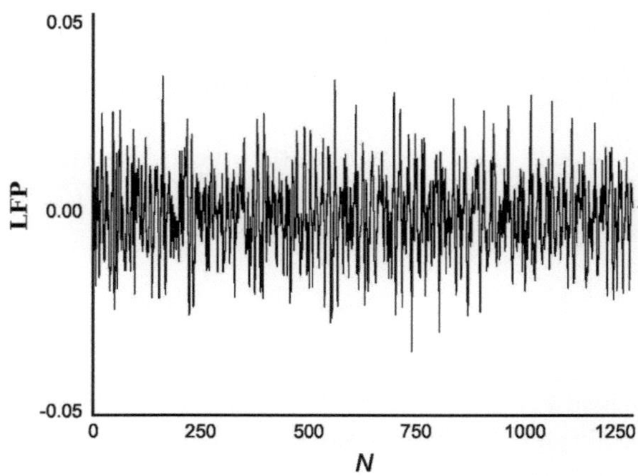

Fig. 1 A representative segment of an LFP signal ($N = 1250$, i.e $t = 12.5$ s) recorded from the ipsi-lateral hemisphere of V1 in a TeNT mice

In particular, for LFP recordings, the relative power content of a specific spectral band λ ($\lambda = \delta, \theta, \alpha, \beta, \gamma$), is a useful quantity having a defined neurophysiological meaning. This measure is defined as follows

$$P_{\text{rel}}(\lambda) = \frac{\sum_{f_k \in \lambda} P(f_k)}{P_{\text{tot}}} \tag{2}$$

where f_k is a frequency belonging to the λ spectral band, $P(f_k)$ is the corresponding power and P_{tot} is the total power of the LFP signal. In Sect. 3.1 we will study the spectral properties of LFP signals recorded from contra-lateral V1 cortex in a focal model of epilepsy in mice.

2.2 *Coupling Measures*

It is well known that distinct brain areas are in communication with each other, however the functional meaning of these interactions is still only partially understood [16]. As an example, in [56], using intracranial EEG signals recorded from epilepsy patients, it was found an influence on the epileptic foci area even from the opposite site of the brain highlighting the complex interactions occurring in different brain areas. Nonlinear time series analysis offers many methods of different nature that provide a quantification of the coupling between neural signals [1, 9, 18, 28, 46]. These methods can be divided in two main classes quantifying the strength of coupling (symmetric measures) and direction of the interaction (asymmetric measures). The former quantifies the overall level of dependence between two neuronal systems, while the latter considers the influence that one system exerts over another. A brief description of some of the most employed methods will be given in the following sections.

2.2.1 Linear and Nonlinear Measures of Coupling Strength

A widely used measure of the linear dependence between two neural signals $x(t_i)$, $y(t_i)$, sampled at $t_i = 0, \Delta t, \ldots, (N-1)\Delta t$, is represented by the cross correlation coefficient [1, 28, 46] defined as

$$\rho_{xy} := \frac{1}{N} \frac{\sum_{i=0}^{N-1} (x(t_i) - \bar{x})(y(t_i) - \bar{y})}{\sigma_x \sigma_y} \tag{3}$$

where $\bar{x}$, $\bar{y}$ and σ_x, σ_y are the mean and the standard deviation of the signal, respectively and N is the total sample size.

On the other hand, if we want to assess the nonlinear correlations between two time series, we can employ the mutual information (MI) measure [1, 28]. A method to estimate the MI between two signals $x(t_i)$ and $y(t_i)$ ($t_i = 0, \Delta t, \ldots, (N-1)\Delta t$), is that of partitioning their supports (X, Y) in a number of bins N_b leading to the following formula:

$$I_{XY} := \sum_{ij} p_{XY}(i,j) \log \frac{p_{XY}(i,j)}{p_X(i) p_Y(j)}, \quad (4)$$

where $p_X(i)$ ($p_Y(j)$) is the probability to find the value of the random variable X (Y) in the ith (jth) bin ($i, j = 1, \ldots, N_b$) and $p_{XY}(i,j)$ is the corresponding joint probability [1, 28, 46]. It is worth remarking that the definition of the cross correlation coefficient is based on second order statistics (linear methods), while the mutual information take into account the full probability distributions (nonlinear methods).

In Sect. 3.2 we will study the coupling strength between LFPs recorded from secondary motor cortices in a model of stroke in mice.

2.2.2 Linear and Nonlinear Measures of Coupling Directionality

In 1969 C. Granger suggested a criterion for testing the causality between time series, say $x(t_i), y(t_i)$ ($t_i = 0, \Delta t, \ldots, (N-1)\Delta t$) [19]. The most common implementation of the Granger causality method is via linear autoregressive modeling of time series data. Using this assumption one can define two measures G_{yx} and G_{xy} quantifying the level of causality from y to x ($y \rightarrow x$) and viceversa ($x \rightarrow y$), respectively. Specifically, G_{yx} is proportional to the ratio σ^x/σ^{yx}, where σ^{yx} represents the variance of the prediction error when the information from y into x is included into the autoregressive model and σ^x is the variance of the prediction error without any information for y (similarly for G_{xy}). Then, the level of causality can be quantified as the ratio

$$R_G = \frac{G_{yx}}{G_{xy}}. \quad (5)$$

According to the above definition, the ratio R_G provides a measure of the directionality of the coupling, i.e. if Y drives X then $R_G > 1$, while $R_G < 1$ if X drives Y. The main limitation of Granger causality resides on the strong assumption that the time series can be modeled by means of linear autoregressive models. To overcome this limitation one can use nonlinear methods [28, 52, 56]. One of the most employed nonlinear methods to assess the coupling directionality between two signals $x(t_i), y(t_i)$ ($t_i = 0, \Delta t, \ldots, (N-1)\Delta t$) is the symbolic transfer entropy [56]. According to [56], the symbolic transfer entropy from $y \rightarrow x$ can be defined as

$$STE_{yx} = \sum p(s_{x,i+\delta}, s_{x,i}, s_{y,i}) \log \frac{p(s_{x,i+\delta}|s_{x,i}, s_{y,i})}{p(s_{x,i+\delta}|s_{x,i})}, \quad (6)$$

where $s_{x,i}$, $s_{y,j}$ are defined through a technique of symbolization and the sum is over all possible symbols (see [56] for details). In the above expression δ is a time step, the logarithm is with base 2, $p(s_{x,i+\delta}, s_{x,i}, s_{y,i})$ is the joint probability of occurrence of the corresponding symbols and $p(s_{x,i+\delta}|s_{x,i}, s_{y,i})$, $p(s_{x,i+\delta}|s_{x,i})$ are conditional probabilities. By changing $s_{x,i}$ with $s_{y,i}$ and vice versa, it is possible to define STE_{xy}, quantifying the coupling level from $x \to y$. Then, a measure of coupling directionality can be defined as

$$R_{STE} = \frac{STE_{yx}}{STE_{xy}}, \tag{7}$$

where $R_{STE} > 1$ if Y drives X and $R_{STE} < 1$ if X drives Y. In Sect. 3.1 we will study the coupling directionality between LFP signals recorded from homologous V1 areas in a model of focal epilepsy in mice.

2.3 *Power Law Correlations*

Physiological recordings in a variety of biological systems are characterized by complex fluctuations exhibiting power-law scaling behavior [4, 6, 63]. The scaling properties of neural signals of humans and animals can be used to get information on the mechanisms modulating neural activities [15, 21, 30, 31, 42, 48]. The existence of scaling implies the absence of characteristic time scales (scale-free) for the physiological signal of interest. Scale-free dynamics is a hallmark of long-range correlation (LRC) that indicates the presence of memory properties in the system. It is well known that an index, called Hurst's exponent ($0 < H < 1$), characterizes the nature of LRC [35]: if $H > 1/2$ the process presents positive LRC, otherwise if $H < 1/2$ the process exhibits anticorrelations.

As pointed out in several research works, spurious identification of LRC can occur [20, 50]. For instance, it has been shown that the presence of trends or periodic components in a signal impacts the performance of the Detrended Fluctuation Analysis (DFA) method [20, 25]. In particular, the existence of anticorrelations in a time series could be strongly biased (i.e. artificial data generated from processes without having LRC could manifest memory properties [50]) and to assess their presence it is convenient to use an integrated approach [50, 58]. In [58], our group has studied the LRC properties of LFP signals by using three different methods simultaneously. Specifically, the Higuchi method [22], the DFA [44] and spectral analysis [28] were employed to estimate the fractal dimension, the Hurst exponent and the scaling properties of the power spectrum, respectively. All these methods consider quantities able to quantify the fluctuations of a signal, specifically the DFA method employs a fluctuation measure $F(n)$ as a function of the considered numbers of points n, the fractal analysis use the length of the fractal curve $\langle L(k) \rangle$ for a fixed time interval k, and the spectral analysis uses the spectral density $S(f)$ as a function of the frequency f. If a scaling exists, then the above measure scales in the following ways:

$$F(n) \sim n^{\alpha_{\mathrm{DFA}}} \qquad H = \alpha_{\mathrm{DFA}} \tag{8}$$

$$\langle L(k) \rangle \sim k^{-D} \qquad H = 2 - D \tag{9}$$

$$S(f) \sim \frac{1}{f^{\beta_s}} \qquad H = \frac{\beta_s + 1}{2} \quad f \to 0, \tag{10}$$

where α_{DFA} is the DFA exponent and it is equal to the Hurst exponent, D is the fractal dimension of the fractal curve and β_s is a coefficient quantifying the scaling behavior of the spectral density. Since the values of these parameters are related to each other and to the Hurst exponent [50, 58], we obtain a consistent scheme for the quantification of the LRC. In Sect. 3.3 we will study the LRC of LFP signals recorded from mice previously reared in EE condition.

3 Characterization of Local Field Potentials in Preclinical Studies of Neural Plasticity

In the following sections, we will present and discuss some results derived from nonlinear time analysis methods applied to LFP signals recorded from different experimental conditions. In particular, we focus on three experimental conditions for animal model (mice) that are relevant for theoretical neuroscience and preclinical studies: Environmental Enrichment (EE) [14, 53, 58], Photothrombotic Stroke (PS) [59], focal model of epilepsy induced by Tetanus neurotoxin (TeNT) [34, 60, 61]. Briefly, EE is an experimental protocol based on a complex sensorimotor stimulation that dramatically affects brain development [14, 53], PS is an induced focal ischemia that reproduces in an animal model the effects of stroke in humans [59] and TeNT, when intracerebrally injected, can induce focal epilepsy in a brain region [34, 60, 61]. It is worth remarking that all the above experimental and computational studies can be considered in the framework of neural plasticity in normal and pathological brain conditions. Neural plasticity is the capacity of the brain to change its basic structures and functions. Neuroplastic changes represent one of the most relevant themes in neuroscience since their comprehension may lead to fundamental advancements in understanding the neurobiological substrates underlying various types of diseases, such as epilepsy or stroke.

Brain areas manifest neuroplastic changes following events like epileptic activity, ischemic injuries or environmental stimulation. In surgical operations, the brain area where seizure originates (“focus”) is often subjected to surgical removal. Unfortunately, sometimes this drastic operation is not enough to stop seizure activity, because another region of the brain can manifest it. This phenomena is called “secondary epileptogenesis” and it represents a crucial example of plasticity [40, 54].

Ischemic infarctions in motor areas are particularly disabling, significantly impacting the execution of many activities of daily life (ADLs). Functional impairment affecting limbs is one of the major challenges that stroke survivors have to face. After an ischemic injury the brain reorganizes spontaneously and limited but significant restoration of motor functions occurs spontaneously after an ischemic damage to the motor cortex. For instance, it has been found that an imbalance of inter-hemispheric communication following stroke could be an obstacle for motor recovery [5, 36, 41].

Another important example in which induced neuroplastic changes are observed is the EE condition [53]. EE is an experimental protocol based on complex sensorimotor stimulations in which animals are keeping free in large cages with an high number of animals and a variety of objects (toys, running wheels, etc...), thus providing an enhancement of sensorimotor stimulation as compared to standard laboratory cage [53]. EE is known to affect several electrophysiological and neuroanatomical parameters in the adult brain. Specifically, EE modulates a series of factors involved in the regulation of synaptic plasticity, including neurotrophic factors, neuromodulators, NMDA receptors, and inhibitory circuitry [53]. Interestingly, it has been found that EE enhanced functional recovery after stroke [26], thus it represents an attractive protocol to be used in post-stroke rehabilitation in humans.

To summarize, from the above discussions on epilepsy, stroke and EE condition we can extrapolate the following questions, i.e.

- What are the neuroplastic changes after stroke?
- What are the neuroplastic changes following epileptic events?
- What are the neuroplastic changes caused by sensorimotor stimulations of an environmental enrichment?
- Do these neuroplastic changes represent true recovery/compensation or maladaptive plasticity?

We believe that answering the above questions will improve the rehabilitation protocol since the functional recovery will be tailored on neurophysiological mechanisms. In fact, a deep understanding of these neurobiological processes will indeed enhance our tools to diagnose, prognosticate, and treat disease. In the following, a particular emphasis will be given to those cases for which the nonlinear time series methods are able to characterize properties that the linear methods fail to capture in a statistically significant way.

3.1 Spectral Properties and Inter-hemispheric Coupling Directionality in a Model of Focal Epilepsy

There is ample clinical and experimental evidence that unilateral epilepsy gradually leads to the generation of a mirror epileptic focus in the homo-topic area of the contra-lateral hemisphere [38, 39, 55]. This finding indicates the occurrence of a strong cross-talk between the hemispheres, which can lead to pathological electrical activity in the contra-lateral side. These considerations clearly indicate the relevance of investigating, using suitable quantitative methods, the properties of the neural activity in the contra-lateral side and coupling directionality between homologous brain regions during epilepsy.

To this aim, in a focal model of epilepsy induced by TeNT injected into the primary visual cortex (V1), we analyzed LFPs recorded from both hemispheres (injected ipsi-lateral and contra-lateral) of V1 in adult mice. LFP signals were acquired in freely moving mice with a sampling rate of $f_s = 100\,\text{Hz}$ as the differential between the two adjacent electrode sites. Each LFP had a total number of points $N = 3 \times 10^5$, 50, 000X amplified and 0.3–30 Hz band-passed [34, 61]. LFP signals were normalized to zero mean and unit variance. The data were obtained from two groups of mice: the TeNT group (number of animals $n_T = 8$) and the control (or vehicle) group (number of animals $n_C = 8$). We were interested in understanding possible modifications of the rhythmic activity in the contra-lateral hemisphere (CL) possibly due to change in the activity of the injected ipsi-lateral (IL) one. To this aim, the power spectrum of each signal was estimated using the Welch method of averaging periodograms, with windows each containing $N_p = 4096$ data points. As specified in Sect. 2.1, the whole power content of each LFP signal was decomposed into normalized frequency bands components (see Eq. 2) corresponding to the classical neurophysiological ones (i.e., $\delta, \theta, \alpha, \beta$). As an example, in Fig. 2 is shown the distribution of these powers among the different bands for the data recorded in the CL hemisphere.

For the CL hemisphere, a statistically significant decrease ($p < 0.01$) of the relative power in the α band was observed. This drop in the α band could be due to changes in the thalamocortical communication (which modulate α activity) after TeNT injection. Indeed, we know that TeNT can reach lateral geniculate nuclei thus impacting on the thalamocortical networks [34]. Furthermore, also a statistically significant increase ($p < 0.05$) of the relative power in the θ band was found. Overall, the above results show that the injection of TeNT determines a redistribution of the relative powers among the corresponding bands occurring mostly in the CL hemisphere. On the other hand, the power content of the spectral bands for the IL hemisphere do not present any statistical differences with respect to the control condition (results not shown).

Moreover, we investigated the inter-hemispheric coupling directionality using the Granger method. This causality test was applied to the pair of LFP signals recorded in the visual cortex of the CL hemisphere and the IL one. In particular, we used a data

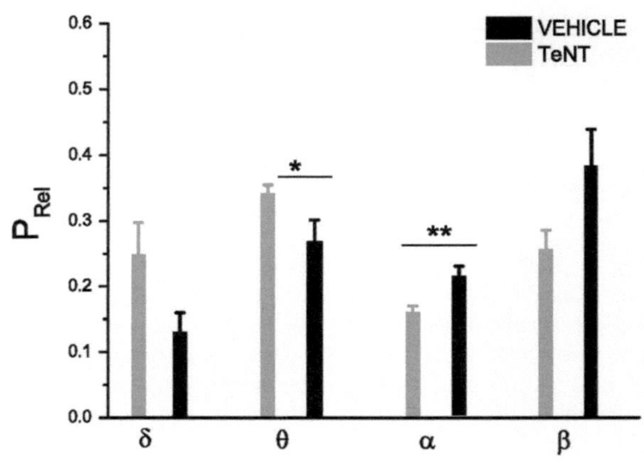

Fig. 2 Mean values and standard errors of the band powers in the contra-lateral hemisphere for TeNT and Vehicle mice. Gray histograms represent TeNT mice, while black histograms refer to Vehicles. A significant decrease of α band was present $p < 0.01$ (One-way Anova), whereas the θ band showed an increase with respect to Vehicle $p < 0.05$ (One-way Anova)

windowing of 512 data points, an autoregressive model of order $p = 15$ (estimated by suitable methods, see [46, 60]) and the average ratio R_G was estimated by averaging over the time windows (see Eq. 5).

The collected values of R_G for all animals in TeNT mice and Vehicle were used to determine the corresponding mean value. With this method, values of $R_G > 1$ indicate that CL hemisphere drives the IL hemisphere, while values of $R_G < 1$ indicate a major influence of the IL side. These boundary conditions for the R_G ratio can be also obtained by using surrogate data. We generated random shuffled data from the LFPs recorded from the control group and we estimated the ratio R_G for each LFP by using a data windowing of 512 data points and autoregressive model of order $p = 15$. Then we averaged the values of the ratios to obtain a mean Granger causality ratio and its corresponding standard error equal to $R^s_G = 1.0002 \pm 0.0001$.

In the left panel of Fig. 3 are reported mean values of R_G for the groups. Although the mean values of TeNT are lower than control, the statistical analysis failed to reveal a significant effect ($p = 0.124$ t-test).

We used the symbolic transfer entropy as a second technique to investigate the coupling directionality between IL and CL hemispheres. In the right panel of Fig. 3 are reported the results obtained with the symbolic transfer entropy. Similarly to the Granger analysis, the surrogate data obtained by random shuffling of the LFPs recorded from Vehicle mice give values for the mean and its corresponding standard errors equal to $R^s_{STE} = 1.01 \pm 0.01$. Moreover, accordingly to Granger analysis the data show a shift of dominance towards the IL, TeNT-injected hemisphere in epileptic mice. Notably, using nonlinear methods we found a statistically significant effect for TeNT mice that was not quantified by Granger analysis. Indeed, the corresponding statistical analysis indicate that the mean values of TeNT and Vehicle are statistically different ($p < 0.05$). These findings suggest rearrangement in the inter-hemispheric coupling after TeNT, probably due to changes in the activity of excitatory pyramidal neurons in the IL hemisphere that mainly mediate the inter-hemispheric connectivity [34, 61].

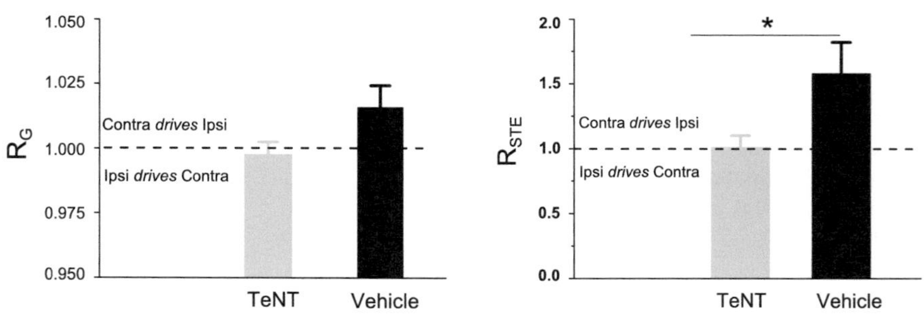

Fig. 3 (Left-panel) Mean values and standard errors of the level of Granger causality ratio R_G for TeNT and Vehicle mice. Gray histograms represent TeNT mice, while black histograms refer to Vehicles. For all panels the error bars represent the corresponding standard errors. The dashed black line corresponds to value equal to one, that distinguishes between the Y(Contra) $>$ X(Ipsi) and X(Ipsi) $>$ Y(Contra) coupling directionality. No significant differences were found between TeNT and Vehicle mice, however a slightly reduction of the Granger causality ratio R_G can be observed. (Right-panel) Same analysis of the left-panel but using Symbolic Transfer Entropy method. We found a decrease of the value of R_{STE} ratio between TeNT and Vehicle mice $p < 0.05$ (t-test)

3.2 Post-stroke Modifications of Inter-hemispheric Coupling in Low Frequency Band

Stroke is one of the main causes of death and disability worldwide [17]. Post-stroke deficits depend on both the location and size of the infarcted brain region. Significant but limited spontaneous restoration of motor function occurs after injury and several studies have aimed at understanding the mechanisms underlying this complex process. In this context, many studies have investigated the interactions between brain hemispheres before and after stroke. Indeed, neural activity in brain motor areas is functionally coupled between the two hemispheres [29] and the lateralization of neural activity during movements is likely to be related to inter-hemispheric inhibition between motor areas exerted via transcallosal connections [10].

In this section, we describe the results obtained from the analysis of LFPs recorded from both hemispheres of either ischemic or control mice. Ischemia was induced unilaterally (right hemisphere) in the Caudal Forelimb Area (CFA), and bipolar electrodes were positioned in both Rostral Forelimb Areas (RFAs) [59]. LFP recordings were collected from freely moving mice at 9 days after stroke (number of animals for the stroke group $n_s = 6$ and $n_C = 6$ for the control group). Each signal consisted of a bivariate time series representing the LFPs simultaneously recorded at sampling frequency $f_s = 200\,\text{Hz}$, i.e. $\Delta t := 1/f_s = 5\,\text{ms}$ and total number of points $N = 3 \times 10^5$ from the ipsi-lesional (right hemisphere) and the contra-lesional hemisphere. Signals were then band-pass filtered in $(0.5-50)\,\text{Hz}$. LFP signals were normalized to zero mean and unit variance.

We investigated the linear and non-linear correlations on signals with frequency components belonging to the δ band ($\delta = (0.5-4)\,\text{Hz}$). To this aim, the signals were filtered to select the requested frequency band and the values of the linear (i.e. cross correlation) and non-linear (i.e. mutual information) correlation were then estimated (see Sect. 2.2.1).

The values of the mutual information for δ band indicated a reduction in stroke mice at day 9 (see left panel of Fig. 4). On the other hand, the cross correlation measure is not able to capture statistically significant differences between the two groups, see right panel of Fig. 4 (although the decreasing trend of correlation is consistent with the mutual information analysis). The present findings suggest possible adaptive inter-hemispheric circuit rearrangement in Pre-Motor Areas which could be the neurophysiological substrate for spontaneous recovery of forelimb motor performance in skilled tasks [59].

3.3 Scaling Properties of LFPs in EE Rearing Condition

Environmental Enrichment (EE) is an experimental protocol where animals are exposed to complex sensorimotor stimulations that affects brain development, including cortico-cortical coupling [14]. Thus, the increased somatosensory stimulation

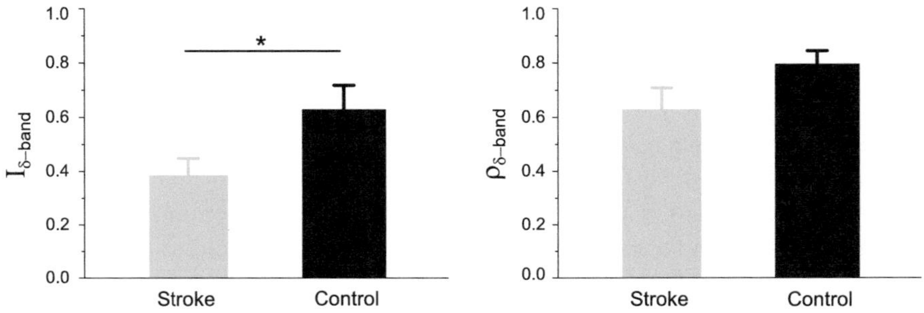

Fig. 4 Analysis of the inter-hemispheric correlation on $\delta = (0.5-4)$ Hz band after stroke. Gray histograms represent stroke mice, while black histograms refer to control mice. For all panels the error bars represent the corresponding standard errors.(Left-panel) Mutual information measure indicated a significant reduction in stroke mice $p < 0.05$ (t-test) with respect to control condition. (Right-panel) Values of the cross-correlation showed no significant changes between stroke and control mice

experienced by an animal raised in EE could impact the dynamical properties of LFPs. For example, in filtered (10 and 20 Hz) EEG signals (amplitude envelopes) from normal human brain, it was shown that LRC are attenuated by somatosensory stimulations [31]. Moreover, the correlation properties of filtered (10 and 20 Hz) MEG and EEG signals (amplitude envelopes) from normal human brain exhibit positive LRC [30]. Although less frequently, also anticorrelations have been observed in physical and biological complex systems [3, 43, 45, 51]. Various theories have been considered to model the scaling behavior of those signals ranging from self-organized critical phenomena [27], filtering properties of the currents through extracellular media [7] and filtering properties of dendrites [37]. Nevertheless, the neural mechanisms responsible for the generation of this type of behavior is still under debate.

Here, we investigated the scaling properties of LFP recordings performed in awake, freely moving mice previously reared in either EE (number of animals $n_{EE} = 6$) and standard condition (SC, number of animals $n_{SC} = 8$) using the protocol described in [14]. Each data set consists of a bivariate time series representing the LFPs simultaneously recorded from visual cortex (V1) and secondary motor cortex (M2). Cortical LFP signals were acquired with a sampling rate of 100 Hz as the differential between two adjacent electrode sites placed in the same cortical area, a total number of points $N = 3 \times 10^5$, 50000X amplified and $(0.3-30)$ HZ band-passed. LFP signals were normalized to zero mean and unit variance.

The LFP recordings have been discovered to present scaling properties that we have quantified using an integrated approach composed by three independent techniques: power spectrum, fractal dimension and DFA [58] (see Sect. 2.3). In particular these signals are characterized by anticorrelations, i.e value of the Hurst exponent $H < 1/2$.

Examples are shown in the panels of Fig. 5, specifically in the top panel is shown the length of the fractal curve $\langle L(k) \rangle$, calculated through the Higuchi method, against the time window k in a double-logarithm scale. For this example the fractal dimension is $D = 1.83$ which is equivalent to an Hurst exponent $H = 2 - D = 0.17$. In the middle panel the results obtained through DFA are shown, in which the scaling of the fluctuation function $F(n)$ is present and the DFA coefficient turned out to be $\alpha = 0.17$ (completely consistent with the Hurst exponent calculated with the fractal dimension D). In the bottom panel the scaling of the power spectrum in the low frequency range ($f < 4$ Hz) is shown. Also in this case the anticorrelation properties are confirmed, i.e. $\beta_s = -0.3$, even if we do not achieve a good agreement with the other above techniques for the precise value of the Hurst exponent (in this case $H = 0.35$).

Notably, the impact of the EE protocol on the scaling properties of the signals is found to be not statistically significant. Indeed, using the t-test and the significance threshold $p = 0.05$ the difference between Hurst's exponents in EE and SC is not statistically significant (see [58] for a complete discussion). This result suggests that the scaling properties of LFP recordings are not affected by the stimulation caused by the EE rearing condition.

These results could have implications in models of LFP, especially in self organized criticality brain models as well as models of scaling due to the extracellular media [7].

4 Discussion and Conclusion

We reviewed some techniques of nonlinear time series analysis applied to neural signals such as LFPs in various experimental conditions. We have shown that nonlinear time series analysis methods are able to quantify effects that are not captured by linear methods, thus highlighting the important role that nonlinear methods can have in studying neurophysiological recordings.

We investigated various properties characterizing the signals such as power spectra, scaling behavior and interdependence (coupling). These characterizations were able to discriminate between different experimental conditions, thus providing a good set of quantities that could be useful as biomarkers in medical diagnostics.

Nevertheless, the biophysical mechanisms underlying the differences in these properties are not well understood. One possibility to improve our knowledge on the neural dynamics of a brain circuit is to build models reproducing the observations. There are various approaches that can be adopted in building models and typically the considered approach depends on the level of knowledge we have on the dynamical system and on what we want to infer from that model.

Models that cannot be recovered starting from first principles, are usually built using the information obtained from the analysis of the data collected during measurements, and they are called phenomenological models [28]. For instance, for neurons these type of models are completely built using the measured biophysical properties of these cells (i.e. detailed description of its parts and their interactions) with parameters having specific physiological meaning. Unfortunately, when we want to

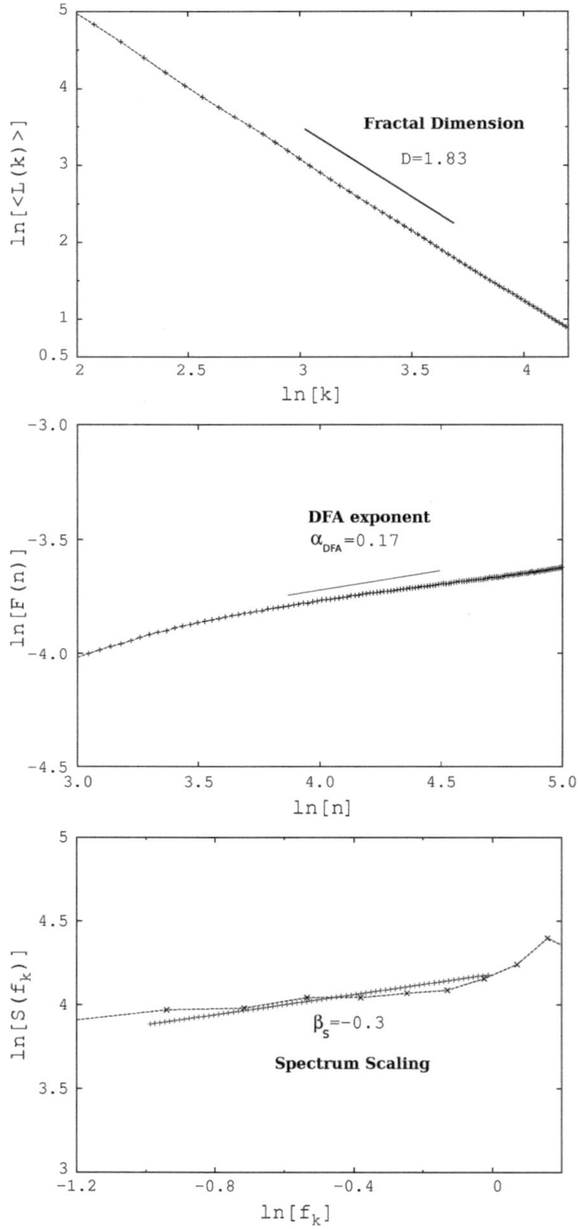

Fig. 5 Integrated approach applied to LFP signal from secondary motor cortex (M2) of a mouse reared in EE. In the top panel is shown the length of the fractal curve, calculated through the Higuchi method, against k in a double-logarithm scale. The fractal dimension turns out to be $D = 1.83$. In the middle panel the results obtained through DFA are shown. The value of the DFA exponent is $\alpha_{\text{DFA}} = 0.17$ and is consistent with the value of Hurst's exponent calculated through the fractal dimension $H_D = 2 - D = 0.17$. In the bottom-panel, the power spectrum $S(f_k)$ of the signal against f_k is shown in a double-logarithm scale (the frequency f_k is expressed in Hz). The corresponding value of β_s is $\beta_s = -0.3$ and the corresponding value of Hurst's exponent is $H_{\beta_s} = \frac{\beta_s+1}{2} = 0.35$

describe theoretically (or computationally) more detailed neural networks composed of hundreds/thousands of coupled neurons there are severe limitation for the intrinsic complexity of the problem.

Another type of approach consists in the modeling of the time series to reproduce the main properties of the data in a statistical sense. If the parameters entering into the model have a physiological meaning, such a model is called parametric [28]. More often, the model equation will have no physiological meaning and the model is called nonparametric [28]. However, even if we cannot recover any information on the physiological mechanisms generating the data this approach turns out to be useful for prediction (also in this case we can have a characterization of the signal with potential implications in medical prognostics).

In general we do not have a recipe to build models and it remains a creative task that probably can be successful when scientists coming from different fields are able to communicate and integrate various expertise.

We think that the recent efforts in understanding brain dynamics will lead to new insights on brain functions and perhaps to new approaches on science with a strong interdisciplinary integration.

References

1. Abarbanel, H.: Analysis of Observed Chaotic Data. Springer Science & Business Media, Berlin (1996). https://books.google.com/books?id=dtTvBwAAQBAJ&pgis=1
2. Antonucci, F., Di Garbo, A., Novelli, E., Manno, I., Sartucci, F., Bozzi, Y., Caleo, M.: Botulinum neurotoxin E (BoNT/E) reduces CA1 neuron loss and granule cell dispersion, with no effects on chronic seizures, in a mouse model of temporal lobe epilepsy. Exp. Neurol. **210**(2), 388–401 (2008). https://doi.org/10.1016/j.expneurol.2007.11.012. http://www.sciencedirect.com/science/article/pii/S0014488607004268
3. Bahar, S., Kantelhardt, J.W., Neiman, A., Rego, H.H.A., Russell, D.F., Wilkens, L., Bunde, A., Moss, F.: Long-range temporal anti-correlations in paddlefish electroreceptors. Exp. Neurol. **210**(3), 454 (2001)
4. Bak, P.: How Nature Works: The Science of Self Organized Criticality. Springer, Berlin (1996)
5. Barry, M.D., Boddington, L.J., Igelström, K.M., Gray, J.P., Shemmell, J., Tseng, K.Y., Oorschot, D.E., Reynolds, J.N.J.: Utility of intracerebral theta burst electrical stimulation to attenuate interhemispheric inhibition and to promote motor recovery after cortical injury in an animal model. Exp. Neurol. **261**, 258–266 (2014). https://doi.org/10.1016/j.expneurol.2014.05.023
6. Bassingthwaighte, J.B., Liebovitch, L.S., West, B.J.: Fractal Physiology. Oxford University Press, New York (1994)
7. Bédard, C., Kröger, H., Destexhe, A.: Does the 1/f frequency scaling of brain signals reflect self-organized critical states? Phys. Rev. Lett. **97**(11), 118,102 (2006). https://doi.org/10.1103/PhysRevLett.97.118102. http://www.ncbi.nlm.nih.gov/pubmed/17025932
8. Beggs, J.: Editorial: can there be a physics of the brain? Phys. Rev. Lett. **114**(22), 220,001 (2015). https://doi.org/10.1103/PhysRevLett.114.220001
9. Bradley, E., Kantz, H.: Nonlinear time-series analysis revisited. Chaos (Woodbury, N.Y.) **25**(9), 097,610 (2015). https://doi.org/10.1063/1.4917289
10. Bütefisch, C., Weßling, M., Netz, J., Seitz, R., Hömberg, V.: Relationship between interhemispheric inhibition and motor cortex excitability in subacute stroke patients. Neurorehabilitation Neural Repair **22**(1), 4–21 (2008)

11. Buzsaki, G.: Rhythms of the Brain. Oxford University Press, Oxford (2006). https://books.google.com/books?id=ldz58irprjYC&pgis=1
12. Buzsáki, G., Anastassiou, C.A., Koch, C.: The origin of extracellular fields and currents EEG, ECoG. LFP and spikes. Nat. Rev. Neurosci. **13**, 407–420 (2012). https://doi.org/10.1038/nrn3241
13. Buzsáki, G., Schomburg, E.W.: What does gamma coherence tell us about inter-regional neural communication? Nat. Neurosci. **18**(4), 484–489 (2015). https://doi.org/10.1038/nn.3952
14. Di Garbo, A., Mainardi, M., Chillemi, S., Maffei, L., Caleo, M.: Environmental enrichment modulates cortico-cortical interactions in the mouse. Plos One **6**(9), 1–11 (2011). https://doi.org/10.1371/journal.pone.0025285
15. Fraiman, D., Chialvo, D.R.: What kind of noise is brain noise? anomalous scaling behavior of the resting brain activity fluctuations. Front. Physiol. **3**(307) (2012)
16. Friston, K., Moran, R., Seth, A.K.: Analysing connectivity with granger causality and dynamic causal modelling. Curr. Opin. Neurobiol. **23**(2), 172–178 (2013). https://doi.org/10.1016/j.conb.2012.11.010. http://www.pubmedcentral.nih.gov/articlerender.fcgi?artid=3925802&tool=pmcentrez&rendertype=abstract
17. Ginsberg, M.D.: Neuroprotection for ischemic stroke: past, present and future. Neuropharmacology **55**(3), 363–389 (2008). https://doi.org/10.1016/j.neuropharm.2007.12.007. http://www.pubmedcentral.nih.gov/articlerender.fcgi?artid=2631228&tool=pmcentrez&rendertype=abstract
18. Gourévitch, B., Bouquin-Jeannès, R.L., Faucon, G.: Linear and nonlinear causality between signals: methods, examples and neurophysiological applications. Biol. Cybern. **95**(4), 349–369 (2006). https://doi.org/10.1007/s00422-006-0098-0
19. Granger, C.W.J.: Investigating causal relations by econometric models and cross-spectral methods. Econometrica **37**(3), 424–438 (1969). https://doi.org/10.2307/1912791
20. Hardstone, R., Poil, S., Schiavone, G., Jansen, R., Nikulin, V.V., Mansvelder, H.D., Linkenkaer-Hansen, K.: Detrended fluctuation analysis: a scale-free view on neuronal oscillations. Front. Physiol. **3**(450) (2012)
21. He, B., Zempel, J., Snyder, A.Z., Raichle, M.E.: The temporal structures and functional significance of scale-free brain activity. Neuron **66**, 353–369 (2010)
22. Higuchi, T.: Physica D **31**, 277–283 (1988)
23. Hlinka, J., Paluš, M., Vejmelka, M., Mantini, D., Corbetta, M.: Functional connectivity in resting-state fMRI: is linear correlation sufficient? NeuroImage **54**(3), 2218–2225 (2011). https://doi.org/10.1016/j.neuroimage.2010.08.042
24. Hodgkin, A.L., Huxley, A.F.: A quantitative description of membrane current and its application to conduction and excitation in nerve. J. Physiol. **117**(4), 500–544 (1952). http://www.pubmedcentral.nih.gov/articlerender.fcgi?artid=1392413&tool=pmcentrez&rendertype=abstract
25. Hu, K., Ivanov, P., Chen, Z., Carpena, P., Stanley, H.: Effect of trends on detrended fluctuation analysis. Phys. Rev. E **64**, 011,114 (2001). https://doi.org/10.1103/PhysRevE.64.011114
26. Janssen, H., Bernhardt, J., Collier, J.M., Sena, E.S., McElduff, P., Attia, J., Pollack, M., Howells, D.W., Nilsson, M., Calford, M.B., Spratt, N.J.: An enriched environment improves sensorimotor function post-ischemic stroke. Neurorehabilitation Neural Repair **24**(9), 802–813 (2010). https://doi.org/10.1177/1545968310372092. http://www.ncbi.nlm.nih.gov/pubmed/20834046
27. Jensen, H.J.: Self-Organized Criticality: Emergent Complex Behavior in Physical and Biological Systems, vol. 1998. Cambridge University Press, Cambridge (1998). https://books.google.fr/books/about/Self&pgis=1
28. Kantz, H., Schreiber, T.: Nonlinear Time Series Analysis. Cambridge University Press, Cambridge (2004). https://books.google.com/books?id=RfQjAG2pKMUC&pgis=1
29. Kinsbourne, M., McMurray, J.: The effect of cerebral dominance on time sharing between speaking and tapping by preschool children. Child Dev. (1974). https://doi.org/10.2307/1128857?uid=35214&uid=3738296&uid=2&uid=3&uid=67&uid=35210&uid=62&sid=21106558162223

30. Linkenkaer-Hansen, K., Nikouline, V., Palva, J., Ilmoniemi, R.: Long-range temporal correlations and scaling behavior in human brain oscillations. J. Neurosci. **21**(4), 1370–1377 (2001)
31. Linkenkaer-Hansen, K., Nikulin, V.V., Palva, J.M., Kaila, K., Ilmoniemi, R.J.: Stimulus-induced change in long-range temporal correlations and scaling behaviour of sensorimotor oscillations. Eur. J. Neurosci. **19**(1), 203–218 (2004). https://doi.org/10.1111/j.1460-9568.2004.03116.x
32. Logothetis, N.K.: What we can do and what we cannot do with fMRI. Nature **453**(7197), 869–878 (2008). https://doi.org/10.1038/nature06976
33. Mainardi, M., Garbo, A.D., Caleo, M., Berardi, N., Sale, A.: Environmental enrichment strengthens corticocortical interactions and reduces amyloid- β oligomers in aged mice. Front. Aging Neurosci. **6**, 1–11 (2014). https://doi.org/10.3389/fnagi.2014.00001
34. Mainardi, M., Pietrasanta, M., Vannini, E., Rossetto, O., Caleo, M.: Tetanus neurotoxin-induced epilepsy in mouse visual cortex. Epilepsia **53**(7), 132–136 (2012). https://doi.org/10.1111/j.1528-1167.2012.03510.x
35. Mandelbrot, B.: The Fractal Geometry of Nature. Times Book, New York (1982)
36. Mansoori, B.K., Jean-Charles, L., Touvykine, B., Liu, A., Quessy, S., Dancause, N.: Acute inactivation of the contralesional hemisphere for longer durations improves recovery after cortical injury. Exp. Neurol. **254**, 18–28 (2014). https://doi.org/10.1016/j.expneurol.2014.01.010
37. Miller, K.J., Sorensen, L.B., Ojemann, J.G., den Nijs, M.: Power-law scaling in the brain surface electric potential. PLoS Comput. Biol. **5**(12), e1000,609 (2009). https://doi.org/10.1371/journal.pcbi.1000609
38. Morrell, F.: Experimental focal epilepsy in animals. AMA Arch. Neurol. **1**(2), 141–147 (1959)
39. Morrell, F.: Cellular pathophysiology of focal epilepsy*. Epilepsia **10**(4), 495–505 (1969)
40. Morrell, F., deToledo Morrell, L.: From mirror focus to secondary epileptogenesis in man: an historical review. Adv. Neurol. **81**, 11–23 (1998)
41. Murase, N., Duque, J., Mazzocchio, R., Cohen, L.G.: Influence of interhemispheric interactions on motor function in chronic stroke. Ann. Neurol. **55**(3), 400–409 (2004). https://doi.org/10.1002/ana.10848. http://www.ncbi.nlm.nih.gov/pubmed/14991818
42. Palva, J.M., Zhigalov, A., Hirvonen, J., Korhonen, O., Linkenkaer-Hansen, K., Palva, S.: Neuronal long-range temporal correlations and avalanche dynamics are correlated with behavioral scaling laws. Proc. Natl. Acad. Sci. **110**(9), 3585–3590 (2013). https://doi.org/10.1073/pnas.1216855110
43. Pan, C., Zheng, B., Wu, Y., Wang, Y., Tang, X.: Detrended fluctuation analysis of human brain electroencephalogram. Phys. Lett. A **329**(12), 130–135 (2004)
44. Peng, C.K., Buldyrev, S.V., Havlin, S., Simons, M., Stanley, H.E., Goldberger, A.L.: Mosaic organization of DNA nucleotides. Phys. Rev. E **49**, 1685–1689 (1994). https://doi.org/10.1103/PhysRevE.49.1685
45. Peng, C.K., Mietus, J., Hausdorff, J.M., Havlin, S., Stanley, H.E., Goldberger, A.L.: Long-range anticorrelations and non-gaussian behavior of the heartbeat. Phys. Rev. Lett. **70**, 1343–1346 (1993). https://doi.org/10.1103/PhysRevLett.70.1343
46. Pereda, E., Quiroga, R.Q., Bhattacharya, J.: Nonlinear multivariate analysis of neurophysiological signals. Prog. Neurobiol. **77**(1), 1–37 (2005). https://doi.org/10.1016/j.pneurobio.2005.10.003
47. Peyrache, A., Dehghani, N., Eskandar, E.N., Madsen, J.R., Anderson, W.S., Donoghue, J.A., Hochberg, L.R., Halgren, E., Cash, S.S., Destexhe, A.: Spatiotemporal dynamics of neocortical excitation and inhibition during human sleep. Proc. Natl. Acad. Sci. U. S. A. **109**(5), 1731–1736 (2012). https://doi.org/10.1073/pnas.1109895109. http://www.pnas.org/content/109/5/1731.abstract
48. Poil, S.S., Hardstone, R., Mansvelder, H.D., Linkenkaer-Hansen, K.: Critical-state dynamics of avalanches and oscillations jointly emerge from balanced excitation/inhibition in neuronal networks. J. Neurosci. **32**(29), 9817–9823 (2012)
49. Rabinovich, M.I., Varona, P., Selverston, A.I., Abarbanel, H.D.I.: Dynamical principles in neuroscience. Rev. Mod. Phys. **78**(4), 1213–1265 (2006). https://doi.org/10.1103/RevModPhys.78.1213

50. Rangarajan, G., Ding, M.: Integrated approach to the assessment of long range correlation in time series data. Phys. Rev. E **61**, 4991–5001 (2000). https://doi.org/10.1103/PhysRevE.61.4991
51. Richards, P.M.: Theory of one-dimensional hopping conductivity and diffusion. Phys. Rev. B **16**, 1393–1409 (1977). https://doi.org/10.1103/PhysRevB.16.1393
52. Romano, M.C., Thiel, M., Kurths, J., Grebogi, C.: Estimation of the direction of the coupling by conditional probabilities of recurrence. Phys. Rev. E Stat. Nonlinear Soft Matter Phys. **76**(3 Pt 2), 036,211 (2007). https://doi.org/10.1103/PhysRevE.76.036211
53. Sale, A., Berardi, N., Maffei, L.: Environment and brain plasticity: towards an endogenous pharmacotherapy. Physiol. Rev. **94**, 189 (2014)
54. , .: Epilepsy as an example of neural plasticity. Neurosci. Rev. J. Bringing Neurobiol. Neurol. Psychiatry **8**(2), 154–173 (2002). http://www.ncbi.nlm.nih.gov/pubmed/11954560
55. Sobayo, T., Fine, A.S., Gunnar, E., Kazlauskas, C., Nicholls, D.: Synchrony dynamics across brain structures in limbic epilepsy vary between initiation and termination phases of seizures. IEEE Trans. Biomed. Eng. **60**(3), 821–829 (2013). https://doi.org/10.1109/TBME.2012.2189113
56. Staniek, M., Lehnertz, K.: Symbolic transfer entropy. Phys. Rev. Lett. **100**(15), 158,101 (2008). https://doi.org/10.1103/PhysRevLett.100.158101
57. Steriade, M.: Neuronal Substrates of Sleep and Epilepsy. Cambridge University Press, Cambridge (2003). https://books.google.com/books?id=J--vNNwDgG8C&pgis=1
58. Vallone, F., Cintio, A., Mainardi, M., Caleo, M., Garbo, A.D.: Existence of anticorrelations for local field potentials recorded from mice reared in standard condition and environmental enrichment. Phys. Rev. E **91**, 012,702 (2015). https://doi.org/10.1103/PhysRevE.91.012702
59. Vallone, F., Lai, S., Spalletti, C., Panarese, A., Alia, C., Micera, S., Caleo, M., Di Garbo, A.: Post-stroke longitudinal alterations of inter-hemispheric correlation and hemispheric dominance in mouse pre-motor cortex. PloS One **11**(1), e0146,858 (2016). https://doi.org/10.1371/journal.pone.0146858
60. Vallone, F., Vannini, E., Cintio, A., Caleo, M., Di Garbo, A.: Time evolution of interhemispheric coupling in a model of focal neocortical epilepsy in mice. Phys. Rev. E **94**(3), 032,409 (2016). https://doi.org/10.1103/PhysRevE.94.032409
61. Vannini, E., Restani, L., Pietrasanta, M., Panarese, A., Mazzoni, A., Rossetto, O., Middei, S., Micera, S., Caleo, M.: Altered sensory processing and dendritic remodeling in hyperexcitable visual cortical networks. Brain Struct. Funct. **221** (2015). https://doi.org/10.1007/s00429-015-1080-1
62. Volz, L.J., Eickhoff, S.B., Pool, E.M., Fink, G.R., Grefkes, C.: NeuroImage differential modulation of motor network connectivity during movements of the upper and lower limbs. NeuroImage **119**, 44–53 (2015). https://doi.org/10.1016/j.neuroimage.2015.05.101
63. West, B.J.: Fractal physiology and the fractional calculus: a perspective. Front. Physiol. **1**(12) (2010)

Functional Cliques in Developmentally Correlated Neural Networks

Stefano Luccioli, Ari Barzilai, Eshel Ben-Jacob, Paolo Bonifazi and Alessandro Torcini

Abstract We consider a sparse random network of excitatory leaky integrate-and-fire neurons with short-term synaptic depression. Furthermore to mimic the dynamics

---

P. Bonifazi and A. Torcini, These authors are joint senior authors on this work.

S. Luccioli (✉) · A. Torcini
CNR - Consiglio Nazionale delle Ricerche - Istituto dei Sistemi Complessi, 50019
Sesto Fiorentino, Italy
e-mail: stefano.luccioli@fi.isc.cnr.it

S. Luccioli
INFN - Istituto Nazionale di Fisica Nucleare - Sezione di Firenze, 50019
Sesto Fiorentino, Italy

S. Luccioli · A. Barzilai · E. Ben-Jacob · P. Bonifazi · A. Torcini
Joint Italian-Israeli Laboratory on Integrative Network Neuroscience, Tel Aviv University,
Ramat Aviv, Israel

A. Barzilai
Department of Neurobiology, George S. Wise Faculty of Life Sciences and Sagol School
of Neuroscience, Tel Aviv University, Ramat Aviv, Israel

E. Ben-Jacob · P. Bonifazi
Beverly and Sackler Faculty of Exact Sciences School of Physics and Astronomy, Tel Aviv
University, Ramat Aviv, Israel

P. Bonifazi
Computational Neuroimaging Lab, BioCruces Health Research Institute,
Hospital Universitario Cruces, Plaza de Cruces, s/n E-48903, Barakaldo, Spain
e-mail: paol.bonifazi@gmail.com

A. Torcini
Laboratoire de Physique Théorique et Modélisation, Université de Cergy-Pontoise - CNRS,
UMR 8089, 95302 Cergy-Pontoise, France
e-mail: alessandro.torcini@u-cergy.fr

A. Torcini
Aix Marseille Univ, Inserm, INMED, Institute de Neurobiologie de la Méditerranée
and INS, Institut de Neurosciences des Systémes, Marseille, France

A. Torcini
Aix-Marseille Université, Université de Toulon, CNRS, CPT, UMR 7332,
13288 Marseille, France

F. Corinto and A. Torcini (eds.), *Nonlinear Dynamics in Computational Neuroscience*, PoliTO Springer Series, https://doi.org/10.1007/978-3-319-71048-8_4

of a brain circuit in its first stages of development we introduce for each neuron correlations among in-degree and out-degree as well as among excitability and the corresponding total degree. We analyze the influence of single neuron stimulation and deletion on the collective dynamics of the network. We show the existence of a small group of neurons capable of controlling and even silencing the bursting activity of the network. These neurons form a *functional clique* since only their activation in a precise order and within specific time windows is capable to ignite population bursts.

1 Introduction

The relationship among brain functional and structural connectivity and neuronal activity is one of the subject of major interest in neuroscience [1–5]. Furthermore, this research line is also strongly related to nonlinear dynamics themes regarding the influence of topology on the emergence of specific microscopic and collective dynamics [6–8].

On the one side, it is nowdays clear that a specific topology is not sufficient to guarantee an unequivocal dynamical behaviour in the neural network [9]. On the other side, in the last two years experimental and numerical evidences have been indicating that neuronal ensembles or cell assemblies (*cliques*) are the emergent functional units of cortical activity [10–13], shaping spontaneous and stimuli/task-evoked responses. Interestingly, several experimental studies have revealed that single neurons can have a relevant role in shaping neuronal dynamics in brain circuits [14–21]. Therefore it is of major importance to understand how do neuronal cliques emerge and how they do operate in relation to single neuron dynamics. In this framework we could link the single neuron firing to the emergence of neuronal cliques, in a sort of hierarchical modular approach where high order dynamics (i.e involving large number of neurons) can be strongly impacted by single neuron manipulations.

Following this line of experimental and theoretical evidence, and in particular the analysis performed in [21], we have derived a numerical model displaying collective oscillations, similar to the ones observed in the hyppocampus and in the neocortex at the early stage of development [22]. In particular, we have analyzed the influence of single neuron stimulation/deletion on the dynamics of such a network in presence of developmentally inspired constraints. Our analysis reveals that a small group of critical neurons, organized in a *clique*, is capable to control the bursting behaviour of the whole network. These neurons are *hubs* in a *functional* sense, because their relevance is not related to high intrinsic connectivity, but to their precise sequential and coordinated activation before a bursting event. As a consequence, if a perturbation is applied to any critical neuron of the *functional clique*, through their stimulation or the deletion, their sequential activation can be interrupted, thus leading to dramatic consequence at a network level, with the arrest of the collective oscillations.

The studied model and the methods employed to analyze the data are introduced in Sect. 2, with particular emphasis on functional connectivity. Section 3 reports the results of the numerical experiments performed at the single neuron level in our network and of the functional analysis. Finally a brief discussion is reported in Sect. 4.

2 Model and Methods

We consider a network of N excitatory Leaky Integrate-and-fire (LIF) neurons, interacting via synaptic currents regulated by short-term-plasticity according to the model introduced in [23]. For this model, the evolution of the membrane potential V_i of the neuron i is given by

$$\tau_{\rm m} \dot{V}_i = -V_i + I_i^{\rm syn} + I_i^{\rm b} \tag{1}$$

where $\tau_{\rm m}$ is the membrane time constant, $I_i^{\rm syn}$ is the synaptic current received by neuron i from all its presynaptic inputs and $I_i^{\rm b}$ represents its level of intrinsic excitability, encompassing single neuron properties as well as excitatory inputs arriving from distal areas.The currents are measured in voltage units (mV), since the membrane input resistance is incorporated into the currents.

Whenever the membrane potential $V_i(t)$ reaches the threshold value $V_{\rm th}$, it is reset to $V_{\rm r}$, and a δ-spike is sent towards the postsynaptic neurons. Accordingly, the spike-train $S_j(t)$ produced by neuron j, is defined as

$$S_j(t) = \sum_m \delta(t - t_j(m)), \tag{2}$$

where $t_j(m)$ represent the m-th spike time emission of neuron j. The transmission of the spike train S_j to the efferent neurons is mediated by the synaptic evolution. In particular, by following [24] the state of the synapse between the jth presynaptic neuron and the ith postsynaptic neuron is described by three adimensional variables, X_{ij}, Y_{ij}, and Z_{ij}, representing the fractions of synaptic resources in the recovered, active, and inactive state, respectively.

Since these three variables should satisfy the constraint $X_{ij} + Y_{ij} + Z_{ij} = 1$, it is sufficient to provide the time evolution for two of them, namely

$$\dot{Y}_{ij} = -\frac{Y_{ij}}{T_{ij}^{I}} + u_{ij} X_{ij} S_j \tag{3}$$

$$\dot{Z}_{ij} = \frac{Y_{ij}}{T_{ij}^{\rm I}} - \frac{Z_{ij}}{T_{ij}^{\rm R}}. \tag{4}$$

Only the active transmitters react to the incoming spikes S_j and the adimensional parameters u_{ij} tunes their effectiveness. The decay times of the postsynaptic current are given by $\{T_{ij}^{\rm I}\}$, while $\{T_{ij}^{\rm R}\}$ represent the recovery times from the synaptic depression.

The synaptic current is expressed as the sum of all the active transmitters (postsynaptic currents) delivered to neuron i

$$I_i^{\text{syn}} = \frac{G_i}{K_i^I} \sum_{j \neq i} \varepsilon_{ij} Y_{ij}, \tag{5}$$

where G_i is the coupling strength, while ε_{ij} is the connectivity matrix whose entries are set equal to 1 (0) if the presynaptic neuron j is connected to (disconnected from) the postsynaptic neuron i. At variance with [23], we assume that the coupling strengths are the same for all the synapses afferent to a certain neuron i.

In particular, we study the case of excitatory coupling between neurons, i.e. $G_i > 0$. Moreover, we consider a sparse network made of $N = 200$ neurons where the i-th neuron has K_i^I (K_i^O) afferent (efferent) synaptic connections distributed as in a directed Erdös–Rényi graph with average in-degree $\bar{K}^I = 10$, as a matter of fact also the average out-degree was $\bar{K}^0 = 10$. The sum appearing in (5) is normalized by the input degree K_i^I to ensure homeostatic synaptic inputs [25, 26].

The intrinsic excitabilities of the single neurons $\{I_i^b\}$ are randomly chosen from a flat distribution of width 0.45 mV centered around the value $V_{\text{th}} = 15\,\text{mV}$, by further imposing that 5% of neurons are above threshold in order to observe a bursting behaviour in the network. For the other parameters, in analogy with Ref. [23], we use the following set of values: $\tau_{\text{m}} = 30\,\text{ms}$, $V_{\text{r}} = 13.5\,\text{mV}$, $V_{\text{th}} = 15\,\text{mV}$. The synaptic parameters $\{T_{ij}^{\text{I}}\}$, $\{T_{ij}^{\text{R}}\}$, $\{u_{ij}\}$ and $\{G_i\}$ are Gaussian distributed with averages $\overline{T^{\text{I}}} = 3\,\text{ms}$, $\overline{T^{\text{R}}} = 800$ ms, $\overline{u} = 0.5$ and $\overline{G} = 45\,\text{mV}$, respectively, and with standard deviation equal to the half of the average.

2.1 Correlations

In this paper, we intend to mimic neural networks at the first stage of development. In such networks mature and young neurons are present at the same time, and this is reflected in the variability of the structural connectivities and of the intrinsic excitabilities. In particular, experimental observations indicate that younger cells have a more pronounced excitability [27, 28], while mature cells exhibit a higher number of synaptic inputs [21, 29]. Thus suggesting that the number of afferent and efferent synaptic connections [21, 29, 30] as well as their level of hyperpolarization [31] are positively correlated with the maturation stage of the cells. Therefore, we consider a network including the following two types of correlations:

- setup T1: a positive correlation between the in-degree K_i^I and out-degree K_i^O of each neuron;
- setup T2: a negative correlation between the intrinsic neuronal excitability and the total connectivity $K_i^T = K_i^I + K_i^O$ (in-degree plus out-degree).

Correlation of type T1 is obtained by generating randomly two pools of N input and output degrees from an Erdös-Rényi distribution with average degree equal to 10. The degrees are ordered within each pool and then assigned to N neurons in order to obtain a positive correlation between K_i^O and K_i^I.

Correlation of type T2 imposes a negative correlation between excitability I_i^b and the total degree of the single neuron $K_i^T = K_i^I + K_i^O$. To generate this kind of correlation the intrinsic excitabilities are randomly generated, as explained above, and then assigned to the various neurons accordingly to their total connectivities K_i^T, thus to ensure an inverse correlation between I_i^b and K_i^T.

2.2 *Functional Connectivity*

In order to highlight statistically significant time-lagged activations of neurons, for every possible neuronal pair, we measure the cross-correlation between their spike time series. On the basis of this cross-correlation we eventually assign a directed functional connection among the two considered neurons, similarly to what reported in [21, 32] for calcium imaging studies.

For every neuron, the action potentials timestamps were first converted into a binary time series with one millisecond time resolution, where ones (zeros) marked the occurrence (absence) of the action potentials. Given the binary time series of two neurons a and b, the cross correlation was then calculated as follows:

$$C_{ab}(\tau) = \frac{\sum_{t=\tau}^{T-\tau} a_{t+\tau} b_t}{min(\sum_{i=1}^{T} a_i, \sum_{k=1}^{T} b_k)} \tag{6}$$

where $\{a_t\}$,$\{b_t\}$ represented the considered time series and T was their total duration. Whenever $C_{ab}(\tau)$ presented a maximum at some finite time value τ_{max} a functional connection was assigned between the two neurons: for $\tau_{max} < 0$ ($\tau_{max} > 0$) directed from a to b (from b to a). A directed functional connection cannot be defined for an uniform cross-correlation corresponding to uncorrelated neurons or for synchronous firing of the two neurons associated to a Gaussian $C_{ab}(\tau)$ centered at zero. To exclude the possibility that the cross correlation could be described by a Gaussian with zero mean or by a uniform distribution we employed both the Student´s t-test and the Kolmogorov-Smirnov test with a level of confidence of 5%. The functional out-degree D_i^O (in-degree D_i^I) of a neuron i corresponded to the number of neurons which were reliably activated after (before) its firing.

For more details on the model and methods employed to perform the reported analysis see [13].

3 Results

In this paper we intend to mimic an immature neuronal network at a certain stage of its initial development, in analogy with the one examined in the experimental work on developmental hippocampal circuits [21]. As discussed in [21, 22], at early postnatal stages such networks are characterized by the excitatory action of GABAergic transmission and the presence of synchronized network events, as largely documented in central and peripheral nervous circuits.

Therefore, we consider a network model composed of only excitatory neurons and displaying bursting activity. A minimal model to mimic the experimentally described stereotypical/characteristic condition of developing neuronal networks [33] is the one introduced by Tsodyks-Uziel-Markram (TUM) [23] with purely excitatory synapses. Since this model (described in details in Sect. 2) is known to display an alternance of short periods of quasi-synchronous firing (*population bursts*, PBs) and long time intervals of asynchronous firing. Furthermore, we consider a network with embedded correlations of type T1 and T2, this in order to account for the presence at the same time of younger and older neurons. This presence can be modeled by considering correlations among the in-degree and out-degree of each cell (type T1) as well as among their intrinsic excitability and connectivity (type T2), as already explained in the previous section. The network activity was characterized by bursts of duration $\simeq$24 ms and with interburst intervals $\simeq$500 ms.

3.1 Single Neuron Stimulation and Deletion Experiments

In the developing hippocampus it has been shown that the stimulation of specific single neurons can drastically reduce the frequency of the PBs [3, 21], or even silence the collective activity. More specifically, the stimulation consisted of current injection capable of inducing sustained high firing regime of the stimulated neuron over a period of a few minutes.

Inspired by this experimental protocol, we test the impact of *single neuron stimulation* (SNS) on the occurrence of PBs in our network model. SNS was achieved by adding abruptly a DC current term to the considered neuron. We report in Fig. 1A–B the stimulation protocol for a specific neuron capable of suppressing the occurrence of PBs for all the duration of the SNS (in this case limited to 4.2 s). The SNS process is totally reversible, i.e. when the stimulation is interrupted the firing rate of the cell and the PBs frequency returns to the pre-stimulation control level.

In order to measure the impact of SNS on the network dynamics, we consider the relative variation of the PB frequency with respect to the control conditions (i.e. in absence of any stimulation). In Fig. 1C–D the impact of a SNS on the PBs frequency is reported for a classical Erdös–Rényi network (no correlations) and a network with embedded correlations T1 plus T2. It is clear that SNS has much more impact on the correlated network, than in the uncorrelated one. In particular for two

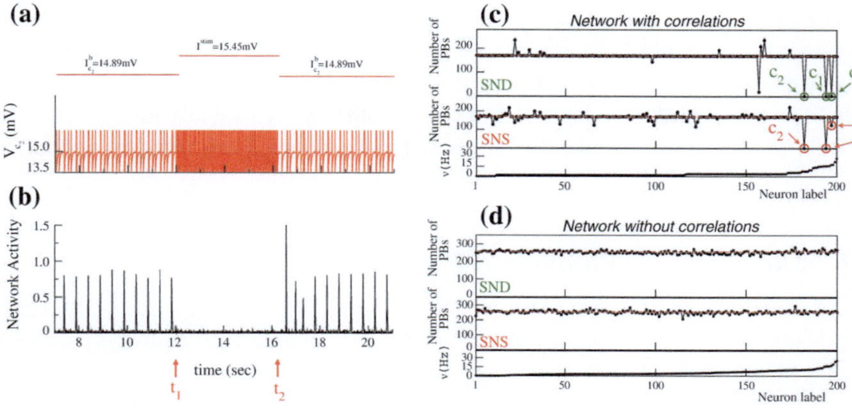

Fig. 1 **A**, **B** Sketch of a SNS experiment for a network with type T1 plus T2 correlations. The neuron c_2 is stimulated with a DC step for a time interval $\Delta t = t_2 - t_1$ (see the red line on the top panel). In **A** the membrane potential V_{c_2} of the stimulated neuron is shown (firing times are signaled by vertical bars of height 1.5 mV), while in **B** the network activity is reported. **C** and **D** refer to correlated and uncorrelated networks, respectively, and display the number of PBs emitted in a time window $\Delta t = 84$ s during SND and SNS experiments with $I^{\text{stim}} = 15.45$ mV. The horizontal dashed lines refer to the average number of PBs emitted in a time interval $\Delta t = 84$ s during a control experiment when no stimulation is applied (the amplitude of the fluctuations is smaller than the symbols). Neurons are ordered accordingly to their average firing rates ν measured in control condition (plotted in the bottom panels). The critical neurons c_0, c_1, c_2, signaled by green (red) circles, are able to strongly affect the bursting during SND (SNS). During SNS experiments each neuron i was stimulated with a DC step switching its excitability from I_i^b to I^{stim} for an interval $\Delta t = 84$ s. The data refer to $I^{\text{stim}} = 15.45$ mV and $N = 200$ neurons

neurons, indicated as neuron c_1 and c_2, SNS performed with stimulation current $I^{\text{stim}} = 15.45$ mV, is able to suppress the occurrence of PBs during the stimulation period. Furthermore in other 7 cases SNS reduces the collective activity $\simeq$30%, in the specific case of neuron c_0 (whose relevance is discussed in the following) the reduction was of $\simeq$27%.

On the contrary, in the network without correlations, the SNS had only marginal effect on the population activity, although the distributions of the firing rates in the correlated and uncorrelated network are extremely similar (under control conditions) as shown in the bottom panels of Fig. 1C and D.

In [23] it has been shown that the elimination of a pool of neurons from an uncorrelated TUM network induced a strong reduction of the population bursts. In our analysis we repeat such numerical experiment with single cell resolution, i.e. we consider the influence of *single neuron deletion*, SND, on the network response. The results reported in Fig. 1C–D, clearly show almost no effect for the uncorrelated network. However, SND blocks the deliver of PBs in the network with correlations $T1$ plus $T2$, in three cases. It should be noticed that two over three neurons were critical also fro the SNS, while the third critical neuron has a higher frequency and it is denoted as c_0.

The most critical neurons are all unable to fire if isolated from the network, i.e. they are all below threshold, apart neuron c_0, which has $I^b_{c_0} = 15.19\,\text{mV}$. Therefore we observe that SND or SNS are capable to silence the network and that this occurs only for a quite limited number of neurons, which have reasonably high I^b but low K^T (c_0, with $K^T = 6$, is ranked as the second less connected neuron of the network).

3.2 The Clique of Functional Hubs

The detailed investigation of the burst events revealed that each PB is always preceded by the firing of the 3 critical neurons in the following order: $c_0 \to c_1 \to c_2 \to$ PB, as it is shown in Fig. 2. The neuron c_0, which is the only one supra-threshold, fires first followed by the others (sub-threshold) and this trigger the onset of the PB. Furthermore, the functional connectivity analysis of the neurons in the network confirms that c_0, c_1, and c_2 have quite high functional output degree $D^O \simeq$170–180 and that they essentially anticipate all the PBs in the network, as shown in Fig. 3B. In more details, one observes that besides firing in a precise order the three critical

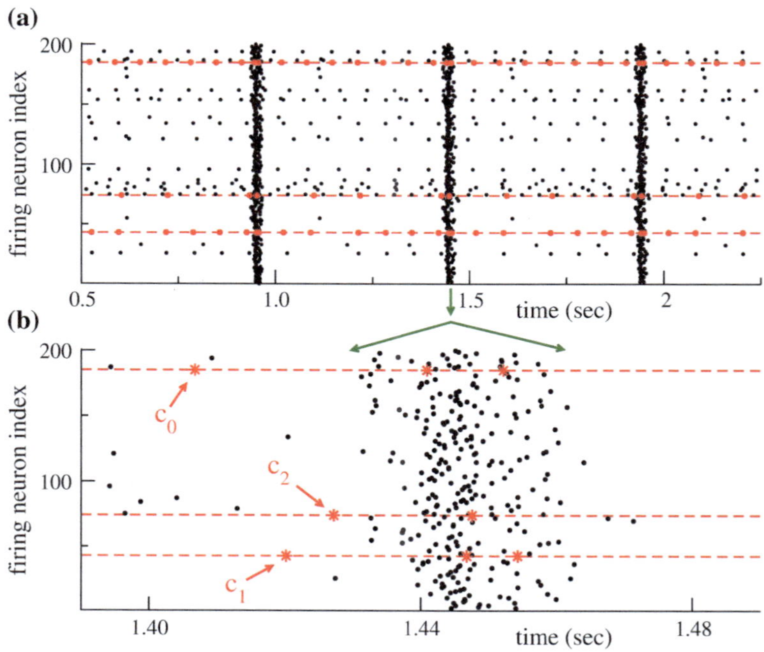

Fig. 2 **A** Raster plot of the network activity: every dot signals a firing event. The (red) dashed lines and (red) stars refer to the critical neurons. **B** Close up of a population burst: PBs are anticipated by the ordered firing sequence of the critical neurons $c_0 \to c_1 \to c_2$. In the raster plots, at variance with all the other figures, the neuronal labels are not ordered accordingly the ascending firing rates

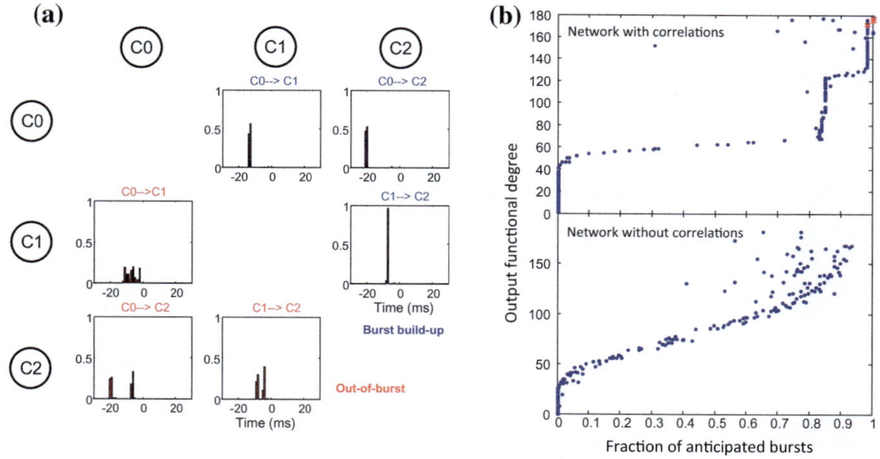

Fig. 3 **A** Cross correlation functions $C(\tau)$ between the spike trains of two critical neurons for the network with correlations. The plots show all the possible pair combinations of the critical neurons: blue (red) histograms refer to the analysis performed during the population burst build up (during periods out of the bursting activity). The order of activation of each pair is reported on the top of the corresponding plot, whenever the cross-correlation has a significant maximum at some finite time τ_{max}. **B** Output functional degree as a function of the fraction of the anticipated population bursts for all the neurons in the network: each dot denotes a different neuron. The data in the top (bottom) panel refer to the network with correlations (without correlations). The (red) stars on the top right in the top panel signal the critical neurons c_0, c_1, c_2

neurons always fire within a narrow temporal window one after the other before a PB as shown in Fig. 3A (in particular see the blue cross correlation functions). Namely, c_1 fires after c_0 with an average time delay $\Delta T_{c_1,c_0} \simeq 13.4$ ms, and then c_2 after c_1 within the average interval $\Delta T_{c_2,c_1} \simeq 7.0$ ms. Furthermore, the analysis of the activity of these three neurons during the inter-bursts reveals that they do not show anymore an univoque firing order or an unique activation time window (see in Fig. 3A the red cross correlations functions).

An extensive investigation of the critical neurons subjected to stimulations with currents in the range $I^{\text{stim}} \in [14.5 : 16.0]$ mV reveals that PBs can be observed only if the neurons c_1 and c_2 had excitabilities within a narrow range (of amplitude $\simeq$0.2 mV) centered around the threshold value. While, the stimulation of neuron c_0 reveals that the network is always active for $I^{\text{stim}} > V_{\text{th}}$, apart the occurence of an absolute minimum in the PB activity (an *anti-resonance*) at $I^{\text{stim}} = 15.32$ mV and a relative minimum at $I^{\text{stim}} = 15.45$ mV (see Fig. 4).

As shown in Fig. 4, whenever c_1 and c_2 are stimulated with currents $I^{stim} > I^b_{c_0} = 15.19$ mV the bursting activity stops. This behaviour indicates that neuron c_0 is the *leader* of the clique and the other ones are simply *followers* in the construction of the PB, they cannot fire more rapidly then neuron c_0 or the PBs ceases. An analysis of the structural connectivity reveales that neuron c_0 projected an efferent synapse on c_1, which projected on c_2.

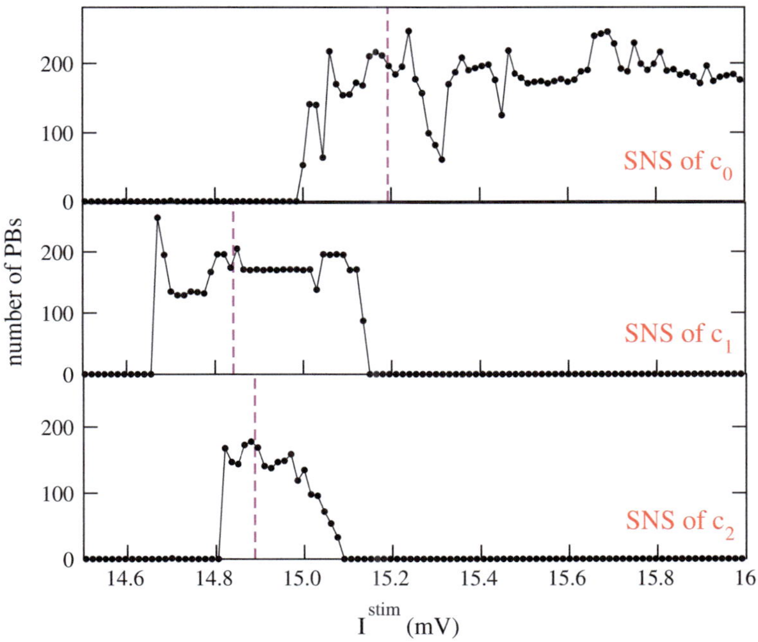

Fig. 4 Number of PBs emitted in a time window $\Delta t = 84$ s during SNS experiments of the critical neurons c_0, c_1, c_2 versus the stimulation current I^{stim}. The vertical dashed lines signal the value of the intrinsic excitability in control condition

The critical neurons, as already mentioned, are not hubs in a structural sense, since they have a very low connectivity K^T, however they are indeed functional hubs, as shown from the previous analysis. Therefore, we can affirm that neurons c_0, c_1 and c_2 form a functional clique, whose sequential activation, within the previously reported time windows, triggers the population burst onset.

The functional relevance of the neurons for the network dynamics is even more evident in other examples of functional cliques, reported in [13], where some of the supra-threshold neurons were even not structurally connected among each other.

4 Discussion

We have shown how, in a simple neural circuit displaying collective oscillations, the inclusion of developmentally inspired correlations among the single neuron excitabilities and connectivities can lead to the emergence of a *functional clique*. This is a small group of functionally connected neurons whose activation can be crucial to promote/arrest the collective firing activity in neural networks, irrespective of the underlying network topology. The clique is composed of a leader neuron, which can start the activation sequence at any moment, but to ignite the population burst the other two neurons, the followers, should fire in a precise order and with quite defined delays.

These results besides being of extreme interest for the neuroscience and dynamical system communities, pave the way for a new approach to the control of the dynamics of neural circuits. Coherence or decoherence in the network activity can be induced by proper stimulation protocols of few peculiar neurons, a subject of extreme interest in particular for the treatment of Parkinson disease with adaptive deep brain stimulation [34].

Future developments towards a more realistic neural circuit would require the extension of the model to include inhibitory neurons, as well as facilitation mechanisms at the level of synaptic transmission, and of network topology arising from anatomical data.

Acknowledgements We thank Y. Ben-Ari, D. Angulo-Garcia, R. Cossart, A. Malvache, L. Módol-Vidal, for extremely useful interactions. This article is part of the research activity of the Advanced Study Group 2016 *From Microscopic to Collective Dynamics in Neural Circuits* performed at Max Planck Institute for the Physics of Complex Systems in Dresden (Germany).

References

1. Friston, K.J.: Functional and effective connectivity in neuroimaging: a synthesis. Hum. Brain Mapp. **2**, 56–78 (1994)
2. Bullmore, E., Sporns, O.: Complex brain networks: graph theoretical analysis of structural and functional systems. Nat. Rev. Neurosci. **10**, 186–198 (2009)
3. Feldt, S., Bonifazi, P., Cossart, R.: Dissecting functional connectivity of neuronal microcircuits: experimental and theoretical insights. Trends Neurosci. **34**, 225–236 (2011)
4. Bullmore, E., Sporns, O.: The economy of brain network organization. Nat. Rev. Neurosci. **13**, 336–349 (2012)
5. Lee, W.C.A., Reid, R.C.: Specificity and randomness: structure-function relationships in neural circuits. Curr. Opin. Neurobiol. **21**, 801–807 (2011)
6. Boccaletti, S., Latora, V., Moreno, Y., Chavez, M., Hwang, D.U.: Complex networks: structure and dynamics. Phys. Rep. **424**(4), 175–308 (2006)
7. Olmi, S., Livi, R., Politi, A., Torcini, A.: Collective oscillations in disordered neural networks. Phys. Rev. E **81**(4), 046119 (2010)
8. Luccioli, S., Olmi, S., Politi, A., Torcini, A.: Collective dynamics in sparse networks. Phys. Rev. Lett. **109**(13), 138103 (2012)
9. Gaiteri, C., Rubin, J.E.: The interaction of intrinsic dynamics and network topology in determining network burst synchrony. Front. Comput. Neurosci. **5**, 10 (2011)
10. Miller, J.E.K., Ayzenshtat, I., Carrillo-Reid, L., Yuste, R.: Visual stimuli recruit intrinsically generated cortical ensembles. Proc. Natl. Acad. Sci. **111**(38), E4053–E4061 (2014)
11. Seabrook, T.A., Huberman, A.D.: Cortical cliques: a few plastic neurons get all the action. Neuron **86**(5), 1113–1116 (2015)
12. Barnes, S.J., Sammons, R.P., Jacobsen, R.I., Mackie, J., Keller, G.B., Keck, T.: Subnetwork-specific homeostatic plasticity in mouse visual cortex in vivo. Neuron **86**(5), 1290–1303 (2015)
13. Luccioli, S., Ben-Jacob, E., Barzilai, A., Bonifazi, P., Torcini, A.: Clique of functional hubs orchestrates population bursts in developmentally regulated neural networks. PLoS Comput. Biol. **10**(9), e1003823 (2014)

14. Tsodyks, M., Kenet, T., Grinvald, A., Arieli, A.: Linking spontaneous activity of single cortical neurons and the underlying functional architecture. Science **286**(5446), 1943–1946 (1999)
15. Brecht, M., Schneider, M., Sakmann, B., Margrie, T.W.: Whisker movements evoked by stimulation of single pyramidal cells in rat motor cortex. Nature **427**, 704–710 (2004)
16. Houweling, A.R., Brecht, M.: Behavioural report of single neuron stimulation in somatosensory cortex. Nature **451**, 65–68 (2007)
17. Cheng-yu, T.L.: Mm, Poo, Dan, Y.: Burst spiking of a single cortical neuron modifies global brain state. Science **324**, 643–646 (2009)
18. Wolfe, J., Houweling, A.R., Brecht, M.: Sparse and powerful cortical spikes. Curr. Opin. Neurobiol. **20**, 306–312 (2010)
19. London, M., Roth, A., Beeren, L., Häusser, M., Latham, P.E.: Sensitivity to perturbations in vivo implies high noise and suggests rate coding in cortex. Nature **466**, 123–127 (2010)
20. Kwan, A.C., Dan, Y.: Dissection of cortical microcircuits by single-neuron stimulation in vivo. Curr. Biol. **22**, 1459–1467 (2012)
21. Bonifazi, P., Goldin, M., Picardo, M.A., Jorquera, I., Cattani, A., et al.: Gabaergic hub neurons orchestrate synchrony in developing hippocampal networks. Science **326**, 1419–1424 (2009)
22. Allène, C., Cattani, A., Ackman, J.B., Bonifazi, P., Aniksztejn, L., et al.: Sequential generation of two distinct synapse-driven network patterns in developing neocortex. J. Neurosci. **28**, 12851–12863 (2008)
23. Tsodyks, M., Uziel, A., Markram, H.: Synchrony generation in recurrent networks with frequency-dependent synapses. J. Neurosci. **20**: 50RC+ (2000)
24. Tsodyks, M.V., Markram, H.: The neural code between neocortical pyramidal neurons depends on neurotransmitter release probability. Proc. Natl. Acad. Sci. **94**, 719–723 (1997)
25. Turrigiano, G.G., Leslie, K.R., Desai, N.S., Rutherford, L.C., Nelson, S.B.: Activity-dependent scaling of quantal amplitude in neocortical neurons. Nature **391**, 892–896 (1998)
26. Turrigiano, G.G.: The self-tuning neuron: synaptic scaling of excitatory synapses. Cell **135**, 422–435 (2008)
27. Ge, S., Goh, E.L., Sailor, K.A., Kitabatake, Y.: Gl, Ming, et al.: Gaba regulates synaptic integration of newly generated neurons in the adult brain. Nature **439**, 589–593 (2005)
28. Doetsch, F., Hen, R.: Young and excitable: the function of new neurons in the adult mammalian brain. Curr. Opin. Neurobiol. **15**, 121–128 (2005)
29. Marissal, T., Bonifazi, P., Picardo, M.A., Nardou, R., Petit, L.F., et al.: Pioneer glutamatergic cells develop into a morpho-functionally distinct population in the juvenile ca3 hippocampus. Nat. Commun. **3**, 1316 (2012)
30. Picardo, M.A., Guigue, P., Bonifazi, P., Batista-Brito, R., Allene, C., et al.: Pioneer gaba cells comprise a subpopulation of hub neurons in the developing hippocampus. Neuron **71**, 695–709 (2011)
31. Karayannis, T., García, N.V.D.M., Fishell, G.J.: Functional adaptation of cortical interneurons to attenuated activity is subtype-specific. Front. Neural Circuits **6**, 66 (2012)
32. Bonifazi, P., Difato, F., Massobrio, P., Breschi, G.L., Pasquale, V., et al.: In vitro large-scale experimental and theoretical studies for the realization of bi-directional brain-prostheses. Front. Neural Circuits **7**, 40 (2013)
33. Ben-Ari, Y.: Excitatory actions of gaba during development: the nature of the nurture. Nat. Rev. Neurosci. **3**, 728–739 (2002)
34. Little, S., et al.: Adaptive deep brain stimulation in advanced Parkinson disease. Ann. Neurol. **74**, 449–457 (2013)

Chimera States in Pulse Coupled Neural Networks: The Influence of Dilution and Noise

Simona Olmi and Alessandro Torcini

Abstract We analyse the possible dynamical states emerging for two symmetrically pulse coupled populations of leaky integrate-and-fire neurons. In particular, we observe broken symmetry states in this set-up: namely, breathing chimeras, where one population is fully synchronized and the other is in a state of partial synchronization (PS) as well as generalized chimera states, where both populations are in PS, but with different levels of synchronization. Symmetric macroscopic states are also present, ranging from quasi-periodic motions, to collective chaos, from splay states to population anti-phase partial synchronization. We then investigate the influence of disorder, such as random link removal or noise, on the dynamics of collective solutions in this model. As a result, we observe that broken symmetry chimera-like states, with both populations partially synchronized, persist up to 80% of broken links and up to noise amplitudes $\simeq$8% of threshold-reset distance. Furthermore, the

S. Olmi
Aix Marseille University, Inserm, INS, Institut de Neurosciences des Systémes, Marseille, France

S. Olmi · A. Torcini (✉)
CNR - Consiglio Nazionale delle Ricerche - Istituto dei Sistemi Complessi, 50019 Sesto Fiorentino, Italy
e-mail: alessandro.torcini@u-cergy.fr

S. Olmi
INFN - Istituto Nazionale di Fisica Nucleare - Sezione di Firenze, 50019 Sesto Fiorentino, Italy
e-mail: simona.olmi@univ-amu.fr

A. Torcini
Laboratoire de Physique Théorique et Modélisation, Université de Cergy-Pontoise - CNRS, UMR 8089, 95302 Cergy-Pontoise, France

A. Torcini
Aix Marseille Univ, Inserm, INMED, Institute de Neurobiologie de la Méditerranée and INS, Institut de Neurosciences des Systémes, Marseille, France

A. Torcini
Aix-Marseille Université, Université de Toulon, CNRS, CPT, UMR 7332, 13288 Marseille, France

F. Corinto and A. Torcini (eds.), *Nonlinear Dynamics in Computational Neuroscience*, PoliTO Springer Series, https://doi.org/10.1007/978-3-319-71048-8_5

introduction of disorder on symmetric chaotic state has a constructive effect, namely to induce the emergence of chimera-like states at intermediate dilution or noise level.

1 Introduction

The emergence of broken symmetry states (*Chimera states*) in populations of oscillators or rotators is an extremely popular and active research field nowadays, in particular after that experimental evidences for the existence of these states have been reported in several contexts ranging from ensembles of mechanical oscillators, to laser dynamics, to populations of chemical oscillators (for a recent review on the subject see [14]). Chimera states in neural systems have been firstly reported by Sakaguchi for a chain of nonlocally coupled Hodgkin-Huxley models with excitatory and inhibitory synaptic coupling [18], while the first evidence of chimeras in models of globally coupled populations has been reported in [12] for leaky integrate-and-fire (LIF) excitatory neurons. More recently, chimeras in LIF networks have analysed in different contexts, ranging from small-network topology [17], to chains of nonlocally coupled LIFs with refractoriness [20], to the emergence of chimeras in a single fully coupled neural population [4]. Globally pulse coupled Winfree models, reproducing θ-neuron dynamics, also support chimera states, ranging from breathing (periodic and quasi-periodic) to chaotic ones [15].

Since the connectivity in the brain is definitely sparse and noise sources cannot be avoided, it is fundamental in order to understand the possible relevance of chimera-like states in neural dynamics to test for the robustness of these solutions to dilution and to the presence of noise. Studies in this direction have been performed mainly for oscillator models [6–8] or excitable systems [19]. In particular, chimera states in random diluted Erdös-Renyi networks have been observed up to a dilution of 8% of the links [6], furthermore it has been shown that noise has not only a washing out effects on chimera solutions, but it can also have a constructive role promoting new dynamical phenomena [7, 19].

In this paper, we focus on the dynamics of two fully pulse coupled populations of excitatory LIF neurons with stronger synaptic coupling among the neurons of the same population and a weaker coupling with those of the other population, similarly to the simplest set-up showing the emergence of chimera states in phase oscillator networks [1]. Furthermore, the neurons are synaptically connected via the transmission of pulses of finite duration. This model for globally coupled systems reveal the emergence of broken symmetry population states, chimera-like, [12], as well as of chimera states even within a single population [4].

Our main aim is to study how the macroscopic solutions, found in the deterministic fully coupled networks, will be modified by considering randomly connected networks of increasing dilution and by adding noise of increasing amplitude to the system. In particular, after having introduced the considered models in Sect. 2, we will report the complete phase diagram for the macroscopic solutions of the fully coupled case in Sect. 3. These solutions vary from chimera-like, to symmetric solutions with complex dynamic ranging from collective quasi-periodic dynamics to macroscopic chaos. Furthermore, we will concentrate on the effect of random dilution and noise on the dynamics of broken symmetry and chaotic states in Sects. 4 and 5. Finally, we will devote Sect. 6 to a brief discussion of the reported results. The algorithm employed to exactly integrate the fully coupled populations is explained in the Appendix.

2 The Model

Firstly we consider two fully coupled networks, each made of N LIF oscillators. Following Ref. [23], the membrane potential $x_j^{(k)}(t)$ of the $j-th$ oscillator ($j = 1, \ldots, N$) of the kth population ($k = 0, 1$) evolves according to the differential equation,

$$\dot{x}_j^{(k)}(t) = a - x_j^{(k)}(t) + g_s E^{(k)}(t) + g_c E^{(1-k)}(t) \tag{1}$$

where $a > 1$ is the suprathreshold input current, while $g_s > 0$ and $g_c > 0$ gauge the self- and, resp., cross-coupling strength of the excitatory interaction. The discharge mechanism operating in real neurons is modeled by assuming that when the membrane potential reaches the threshold value $x_j^{(k)} = x_{th} = 1$, it is reset to the value $x_j^{(k)} = x_R = 0$, while a α-pulse $p(t) = \alpha^2 t \exp -\alpha t$ is transmitted to and instantaneously received by the connected neurons. For this kind of pulses the field $E^{(k)}(t)$ generated by the neurons of the population k, satisfies the differential equation

$$\ddot{E}^{(k)}(t) + 2\alpha \dot{E}^{(k)}(t) + \alpha^2 E^{(k)}(t) = \frac{\alpha^2}{N} \sum_{j,n} \delta(t - t_{j,n}^{(k)}) \,, \tag{2}$$

where $t_{j,n}^{(k)}$ is the nth spiking time of the jth neuron within the population k, and the sum is restricted to times smaller than t. In the limit case $g_s = g_c = g$, the two populations can be seen as a single one made of $2N$ neurons with an effective coupling constant $G = 2g$.

Secondly we consider two random undirected Erdös-Renyi networks, each made of N LIF oscillators and with an average in-degree K, therefore the probability to have a link between two neurons is simply K/N. We assume that the membrane potential $x_j^{(k)}(t)$ of the $j-th$ oscillator of the kth population ($k = 0, 1$) evolves according to the differential equation

$$\dot{x}_j^{(k)}(t) = a - x_j^{(k)}(t) + g_s E_j^{(k)}(t) + g_c \overline{E}^{(1-k)}(t) \quad ; \tag{3}$$

where the field $E_j^{(k)}(t)$ takes into account the pulses received by the neuron j from the neurons of its own population, while the field $\overline{E}^{(1-k)}(t)$ represents the effect of the neurons belonging to the other population.

In particular, $E_j^{(k)}(t)$ is the linear superposition of the pulses $p(t)$ received by neuron i of the kth-population at all times $t_n < t$ (the integer index n orders the sequence of the pulses emitted in the network), namely:

$$E_j^{(k)}(t) = \frac{1}{K} \sum_i \sum_{n|t_n<t} C_{i(n),j}^{(k)} \theta(t - t_n) p(t - t_n) \quad , \tag{4}$$

where $\theta(x)$ is the Heavyside function and the connectivity matrix $C_{i,j}^{(k)}$ has entries 1 (0) depending if the neuron j presents a post-synaptic neuron connection with neuron i or not. For each neuron we should introduce a different field, since each neuron has a different connectivity in the network. It is more convenient to turn also this time, as previously done for the globally coupled case, the explicit Eq. (4) into the following differential equation

$$\ddot{E}_j^{(k)}(t) + 2\alpha \dot{E}_j^{(k)}(t) + \alpha^2 E_j^{(k)}(t) = \frac{\alpha^2}{K} \sum_{i,n} C_{i(n),j}^{(k)} \delta(t - t_{i,n}^{(k)}) \,. \tag{5}$$

Furthermore, $\overline{E}^{(1-k)}(t) = \frac{1}{N} \sum_{i=1}^{N} E_i^{(1-k)}(t)$ represents a "mean field" effect of the second population on the neuron of the first population, since it is the average of all the fields $E_i^{(1-k)}$ of the second population. As a result, the dynamics of the neural network model takes the more "canonical" form of a set of coupled ordinary differential Eqs. (3) and (5), which can be analyzed with the standard methods of dynamical systems. The setup we have employed, diluted random connectivity within each population, but mean-field like cross coupling, will favour the stabilization of the broken symmetry state as suggested in [6]. The have studied the diluted networks in so-called massively connected case, namely where the average in-degree is proportional to the system size $K = (1 - d) \times N$.

Finally we consider two diluted networks with noise. The noise is introduced in the system every time the membrane potential has reached the threshold value and it is reset to the reset value. In particular, instead of using a reset value $x_R = 0$, the neuron is reset to a random value chosen in the interval $x_R \in [-\Delta, \Delta]$, where Δ takes into account the level of the noise. In this case the percentage of dilution is kept fixed ($d = 0.2$).

The integration of the above models is performed exactly in terms of so-called event driven maps analogously to what previously done in [12, 23], for the two fully coupled cases, where non trivial round-off problems can occur. To avoid these problems a more refined event driven map has been developed and it is explained in the Appendix.

The degree of synchronization within each population of neurons can be quantified by introducing the typical order parameter used for phase oscillators $r^{(k)}(t) = |\langle e^{i\theta_j^{(k)}(t)} \rangle|$, where $\theta_j^{(k)}$ is the phase of the jth oscillator, that can be properly defined as a (suitably scaled) time variable [22], $\theta_j^{(k)}(t) = 2\pi(t - t_{j,n}^{(k)})/(t - t_{m,n-1}^{(k)})$, where n identifies the time of the last spike emitted by the jth neuron, while m identifies the neuron that has emitted the last spike at time t. One can verify that this phase is bounded between 0 and 2π, as it should. Furthermore, the fully synchronized regime corresponds to $r^{(k)} \equiv 1$, and in the asynchronous regime one expects $r^{(k)} \simeq 1/\sqrt{K}$, where K is the average in-degree of the network.

3 Fully Coupled Network: Phase Diagram

The phase plane (g_c, g_s) shown in Fig. 1 has been obtained by studying the model (1,2) for $a = 1.3$ and $\alpha = 9$. As already mentioned, along the diagonal ($g = g_s = g_c$) the two population model (1) reduces to a single population with coupling strength $G = 2g$. For our choice of a and α values, the system exhibits *Partial Synchronization* PS, where the macroscopic field displays collective periodic oscillations and the microscopic dynamics is quasiperiodic [21, 23]. Below the diagonal, the evolution is still symmetric but the neurons are now *Fully Synchronized* (FS); the neurons of both populations fire at the unison. More interesting phenomena can be observed above the diagonal. In this situation a solution with broken symmetry emerges naturally, where one population is FS and the other is PS, this represents a generalized form of chimera state. In particular, one observes that while the order parameter of one population is exactly one, the other oscillates periodically, as shown in Fig. 2c. Therefore this chimera state can be classified as a *periodically breathing chimera*, which has been previously reported for the Kuramoto model [1, 16] as well as for a two population network of rotators in [13]. Despite the macroscopic fields $E^{(0)}$ and E^1 are both oscillating periodically and locked, as evident from Fig. 2a, e, the two populations are characterized by different behaviour at a microscopic level, where the neurons are periodic in the FS population and quasi-periodic in the PS population. This means that the neurons subject to two different linear combinations of $E^{(0)}$ and $E^{(1)}$ behave differently: a population locks with the forcing field, while the other one behaves quasi-periodically.

Another even more interesting symmetry broken state (termed PS1-PS2) can be observed for larger g_s-values and $g_c < 0.055$; in this case both populations exhibit PS, but their dynamics take place over two different attractors with two different degrees of synchronization, as shown in Fig. 2b, d. Analogously to the PS-FS state, the two fields are periodic and phase locked, as it can be seen by looking at the time behavior of $E^{(0)}$ and $E^{(1)}$ in Fig. 2f. However, at variance with PS-FS, here both populations exhibit quasi-periodic motions. This simmetry broken state can be also considered a Chimera state and it has been reported only for LIF populations so-far [12].

For larger g_s values the symmetry between the two collective fields is recovered with the only difference of phase shift between the two fields which oscillate in antiphase and this is why we term this regime *Antiphase Partial Synchronization* (APS) (see Fig. 3a, g). In this regime, at finite N the instantaneous maximum Lyapunov exponent strongly fluctuates and we cannot exclude that this regime is *weakly chaotic*. Analogously to the chaotic behaviour found in single population of

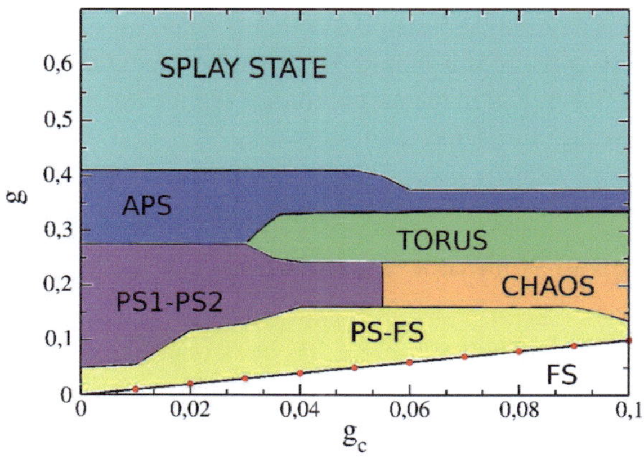

Fig. 1 Phase diagram in the (g_c, g_s)-plane reporting the stability region of the observed various collective solutions. For the definitions of the different phases see the text

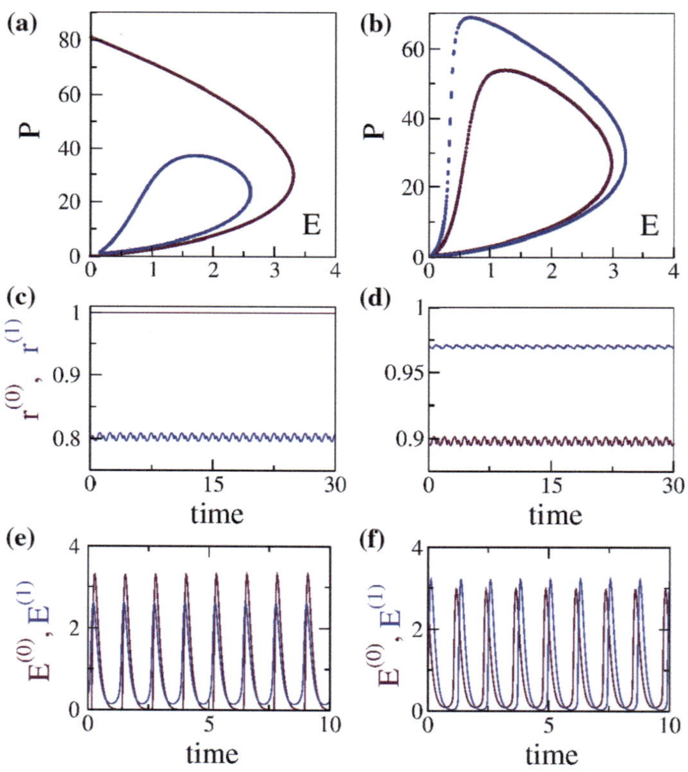

Fig. 2 Macroscopic attractors displayed by reporting $P \equiv E + \alpha \dot{E}$ versus E for a PS-FS state (**a**) and a PS1-PS2 (**b**), the time evolution of the corresponding order parameters $r^{(0)}$ and $r^{(1)}$ is also reported in **c** and **d**. In panels **e**, **f** are reported the time behaviors of the macroscopic fields $E^{(0)}$ and $E^{(1)}$. The variables corresponding to population 0 (resp. 1) are shown in blue (resp. maroon). As regards the parameter values, $(g_c = 0.07, g_s = 0.1)$ in **a**, **c** and $(g_c = 0.02, g_s = 0.17)$ in **b**, **d**

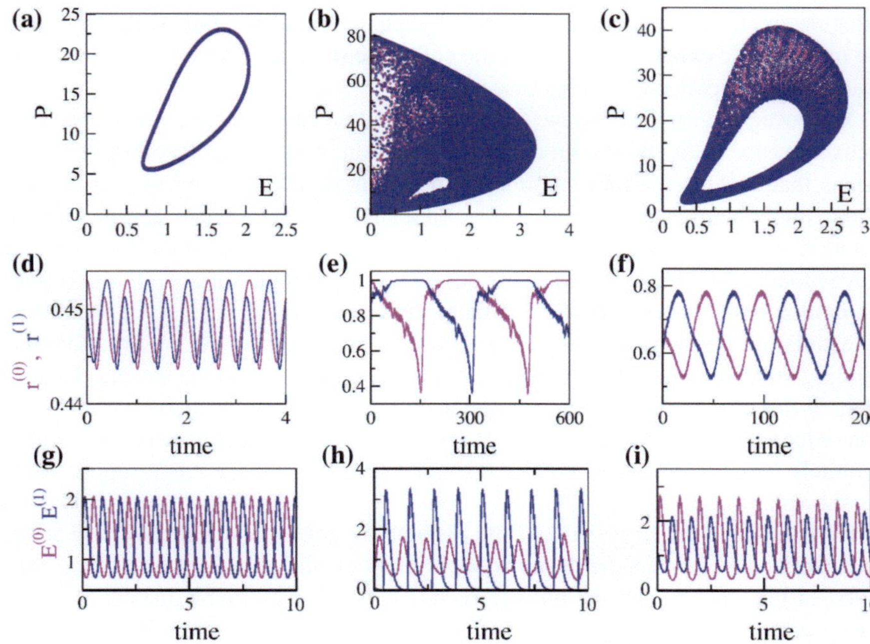

Fig. 3 Macroscopic attractors displayed by reporting $P \equiv E + \alpha \dot{E}$ versus E for an APS (**a**), a chaotic state (**b**), and a TORUS state (**c**), the time evolution of the corresponding order parameters $r^{(0)}$ and $r^{(1)}$ is also reported in **d**, **e** and **f**. In panels **g**, **h**, **i** are reported the time behaviors of the macroscopic fields $E^{(0)}$ and $E^{(1)}$. The variables corresponding to population 0 (resp. 1) are shown in magenta (resp. blue). As regards the parameter values, ($g_c = 0.07$, $g_s = 0.35$) in **a**, **d**, **g**, ($g_c = 0.08$, $g_s = 0.16$) in **b**, **e**, **h** and ($g_c = 0.07$, $g_s = 0.3$) in **c**, **f**, **i**

massively coupled LIFs [11], we expect that the chaoticity disappears in the thermodynamic limit. However, it is peculiar the behaviour of the order parameters in this case, as shown in Fig. 3d: the two populations are not equally synchronized and the two order parameters $r^{(0)}$ and $r^{(1)}$ are behaving periodically in time, but at each oscillation the role of most synchronized population switches from one to the other.

In a limited region above the diagonal and for $g_c > 0.055$ the collective behaviour is still symmetric but irregular (*Collective Chaos*), as revelead by the two macroscopic attractors (see Fig. 3b). Furthermore, in this case one can observe quite wide oscillations of the order parameters of the two populations in the range $0.4 \leq r^{(0)}, r^{(1)} \leq 1$, as shown in Fig. 3c. Whenever one population gets synchronized, with an order parameter $\simeq 1$, the other partially desynchronizes reaching values $r \simeq 0.4$. These collective oscillations in the order parameters occur on quite long timescale with respect to the periods of oscillations of the two macroscopic fields $E^{(0)}$ and $E^{(1)}$ reported in Fig. 3h. However, the oscillations in the level of synchronization induce modulations with periods of the same order in the field dynamics. In a previous paper [12] we have demonstrated that the finite-amplitude Lyapunov exponent [3], for this state, coincides with the microscopic maximal Lyapunov exponent, thus suggesting that the microscopic chaos is induced by the collective drive and therefore the origin of chaos is indeed collective in this case.

Moreover, in a strip above the chaotic region, one can observe a symmetric collective quasiperiodic motion on a *Torus* T^2 for both populations (see Fig. 3c). This means that the quasiperiodic motion of the fields is accompanied by a dynamics of the single neurons along a torus T^3. An analogous regime has been previously reported in [10] for a population of coupled Stuart–Landau oscillators. Here, we find it in a model where the single units are described by an single variable. Furthermore, the motion on the macroscopic T^2 attractor reported in Fig. 3c can be characterized by estimating the winding numbers for various system sizes, we observe that the winding number is constant, indicating that the torus survives in the thermodynamic limit. In this case, we observe quite regular antiphase oscillations in the synchronization order parameters between values $0.4 \le r^{(0)}, r^{(1)} \le 0.8$ occurring on time scales definitely longer than those associated to the oscillations of the macroscopic fields (as shown in Fig. 3f, i).

Finally, for yet larger g_s-values both populations converge towards a *Splay State*, characterized by constant fields, no collective motion and periodic microscopic evolution of the neurons. This is not surprising, as we already know that for the chosen α- and a-values, the splay state is stable in a single population of neurons for $G > G_0 \equiv 0.425$ [21, 23].

4 Diluted Networks

In order to observe the influence of dilution on the dynamics, we considered in absence of dilution a PS-FS state with broken symmetry and a Chaotic state. In particular, we have analyzed the modifications of the macroscopic attractors, of the level of synchronization, as well as the microscopic dynamics induced by cutting randomly links for these 2 states.

To characterize the two macroscopic states, we have decided to consider the level of synchronization in the two populations. In general, we observe that the effect of dilution is, as expected, to reduce the level of synchronization in the system. In particular, starting from the PS-FS state in the fully coupled case, the average values $\bar{r}^{(0)}$ and $\bar{r}^{(1)}$ remain distinct up to some critical dilution $d_c \simeq 0.75$ (as shown in Fig. 5a. For intermediate values of the dilution, in the range $0.2 \le d \le d_c$, a broken symmetry state is still observable, characterized by two periodically oscillating fields with associated different attractors (see Fig. 4a, e, o). Therefore, we can safely classify this as a chimera PS1-PS2 state, despite the dilution induces fluctuations in the macroscopic fields [2, 11]. For larger dilution, above the critical value, the two attractors essentially merge (as shown in Fig. 4i), but both the macroscopic fields are still presenting clear collective periodic oscillations even at these levels of dilution (see Fig. 4c), confirming the robustness of the PS states in this model.

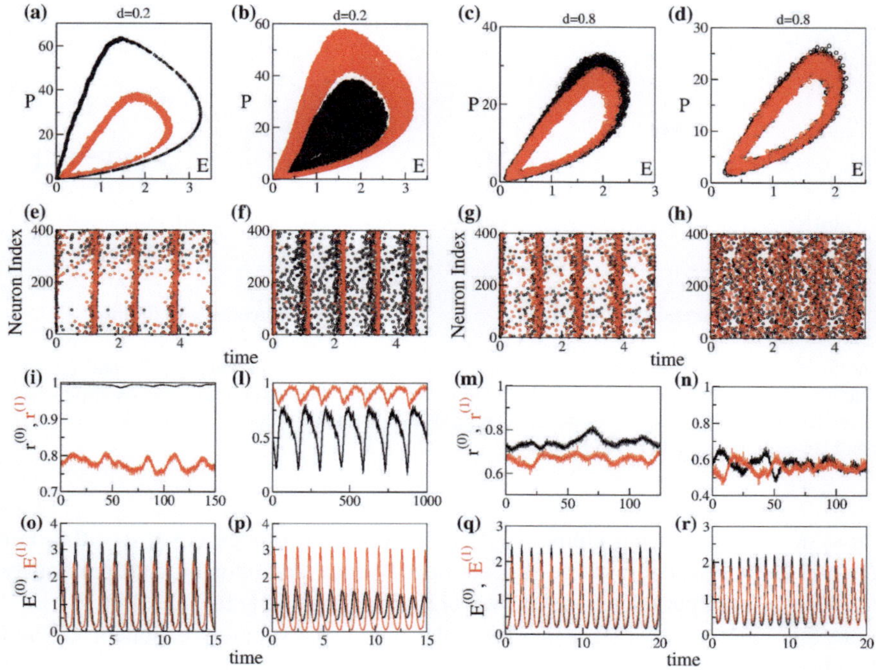

Fig. 4 **Influence of Dilution**. Macroscopic and microscopic characterization for two different dilutions (namely $d = 0.2$ and 0.8) of two states that at $d = 0$ were FS-PS (first and third columns) and chaotic (second and fourth columns). In the first row the corresponding macroscopic attractors are reported, namely, $P \equiv E + \alpha \dot{E}$ versus E are shown; the raster plots are shown in the second row; the time evolution of the order parameters $r^{(0)}$ and $r^{(1)}$ is reported in the third row, while that of the macroscopic fields $E^{(0)}$ and $E^{(1)}$ is shown in the forth row. The variables corresponding to population 0 (resp. 1) are shown in black (resp. red). As regards the parameter values, ($g_c = 0.04$, $g_s = 0.1$) for the first and third columns, and ($g_c = 0.08$, $g_s = 0.16$) for the second and fourth columns. The employed values of dilution are reported over the corresponding columns

As a general remark when we considered the influence of dilution on a PS1-PS2 state, we observed a similar scenario, obviously without an initial window where the FS was still observable.

The dilution has a quite peculiar effect on the chaotic, symmetric, state; in fact, up to dilution $d \simeq 0.2$, we did not observe any new effect, as evidenced by the average value of the synchronization order parameters reported in Fig. 5b. However, already at $d = 0.2$ the dilution induce a symmetry break among the two population dynamics. This is clear in Fig. 4b, where one population is still in a collective chaotic state, similar to the one observed for the globally coupled system, while the other reveals an attractor analogous to the one seen for the PS state. This is even more evident by considering the time evolution of the order parameters, while one population exhibits large oscillations of $r^{(0)}$, similar to the one observed for the chaotic state, the other reveals more limited oscillations (see Fig. 4l). By further increasing the dilution,

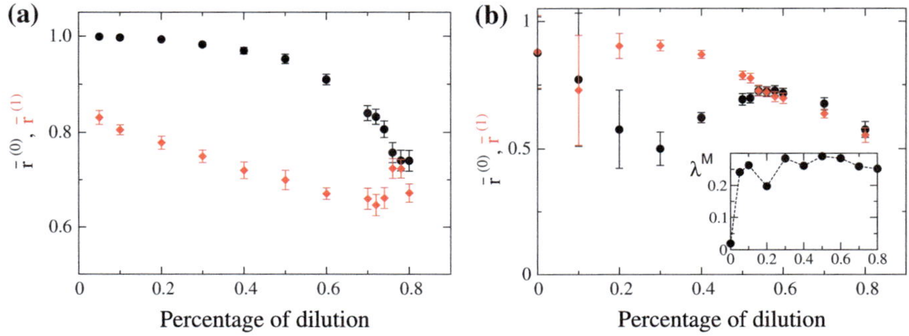

Fig. 5 In panels **a**, **b** are reported the average values of the order parameters $\bar{r}^{(0)}$ and $\bar{r}^{(1)}$ as a function of the percentage of dilution. In the inset is reported the maximum Lyapunov exponent as a function of the percentage of dilution. As regards the parameter values, $(g_c = 0.04, g_s = 0.1)$ in **a** and $(g_c = 0.08, g_s = 0.16)$ in **b**

the system shows a clear chimera PS1-PS2 state over a range $0.3 \le d \le 0.5$. For $d > 0.5$ the two attractors merge in a commons PS state, analogously to the previously considered set of parameters (as shown in Fig. 4d, h, p and r) for $d = 0.8$).

In this latter case we also measured the maximal Lyapunov exponent λ^M and we observed that it stays positive for all the considered dilution values (see inset Fig. 4b). However, while for vanishing dilutions the origin of the chaotic dynamics can be considered as a collective effect induced by the chaotic motion of the coupled macroscopic fields, analogous to the chaotic state observed for two fully coupled populations [12], for larger dilution we expect chaotic effects to be present at the level of the single populations, in the form of (microscopic) *weak chaos*. This form of chaos disappears in the thermodynamic limit, and it is due to stochastic fluctuations of the single macroscopic fields induced by finite in-degree effects [2, 11]. At intermediate dilution both effects are present and the level of chaoticity is bigger with respect to the fully coupled case (where, $\lambda_M \simeq 0.02$); this is also evident from Fig. 4b where one attractor appears as being chaotic, while the other is in a PS state plus finite size fluctuations. To summarize, the system in absence of dilution, thanks to the interaction of the two populations, exhibits *collective chaos*, the dilution induces another form of chaos termed *weak chaos*, because it is present only in systems of finite size. However, for the chosen system size and parameters the level of chaoticity due to finite size fluctuations is definitely higher than that due to collective chaos.

5 Noisy Dynamics

In this case, we consider as unperturbed state a network with a small level of dilution $d = 0.2$ and we study how the noise modify the original dynamics. In absence of noise we consider once more a chimera PS-FS state and a chaotic state. Please, notice that the small dilution modifies the phase diagram shown in Fig. 1 for the globally coupled case. In particular in order to observe a chaotic symmetric state, we have been forced to employ parameter values slightly different from those considered in the previous Section for the same state, namely we used ($g_c = 0.08$, $g_s = 0.2$).

For the chimera FS-PS, it is evident from Fig. 7a that the complete synchronization in one population persists only up to noise of amplitudes $\Delta \simeq 0.02$, however the two populations behave differently over a quite wide range of noise amplitudes (namely, $0 \leq \Delta \leq 0.07$). In all this range we observe chimera states of the type PS1-PS2, obviously with fluctuations in the macroscopic variables induced by noise, as it is evident from Fig. 6a, e, i and o. By further increasing the noise amplitude a complete symmetry is recovered but the fields still exhibit periodic collective oscillations as shown in Fig. 6q. Also all the other indicators suggest that each population is still PS, in particular the synchronization degree remains quite high $\bar{r}^{(0)}, \bar{r}^{(1)} \simeq 0.8$ (see also Fig. 6c, g, m and q).

For what concerns the chaotic state, this remains symmetric and characterized by an unique chaotic attractor up to noises of quite large amplitude, namely $\Delta = 0.05$. As evident, from Fig. 6b, f, l and p, all the characteristics of a collectively chaotic state seem present: overlapping chaotic attractors filling a closed portion of the phase space, anti-phase irregular oscillations in the order parameters over long time scales etc. The quite unexpected result is that by further increasing noise the symmetry of the attractors is broken and the system evolves towards a chimera PS1-PS2 state, which is observable in the range $0.06 \leq \Delta \leq 0.08$ (as shown in Fig. 7b). A specific example of this broken symmetry state is reported in Fig. 6d, h, n and r for $\Delta = 0.08$. For even larger noise amplitudes the two attractors converge towards a common PS state with level of average synchronization $\bar{r} \simeq 0.6$.

6 Discussion

A first important aspect to notice, is that in the present model the Chimera states FS-PS and PS1-PS2 do not coexist with a stable regime where both populations are FS, as usual for Chimera states emerging in phase oscillator populations. This implies that in the present case the initial conditions should not be prepared in some peculiar way to observe the emergence of broken symmetry states, therefore they are not induced by the choice of the initial conditions as in most of the examined models. Spontaneously emerging Chimera, in system where the FS was unstable, have been reported also for chains of Hodgkin-Huxley neurons [18] and of Stuart–Landau oscillators [5].

As general results, we observe that dilution or noise have a similar influence on the studied macroscopic dynamics, despite random dilution represents a quenched form of disorder, while dynamical noise an annealed one. In particular, starting from a broken symmetry state dilution or noise reduce the level of synchronization in the two population, leading the dynamics of the two networks to be more and more similar for increasing dilution/noise. On the other hand, starting from a symmetric state, namely a chaotic one, the role of disorder is to break (at some intermediate dilution or noise

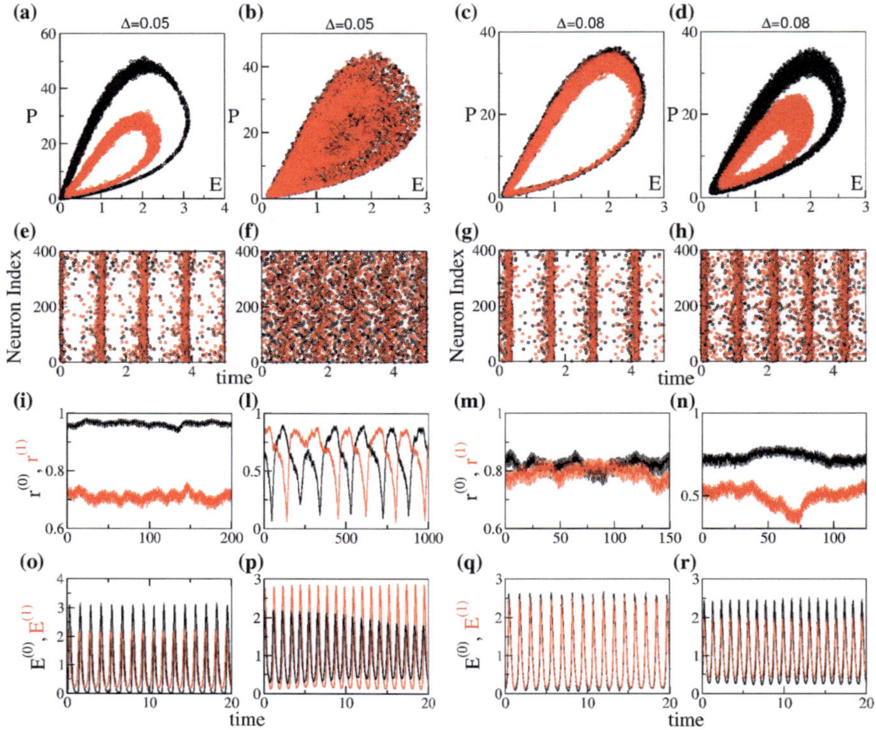

Fig. 6 **Noise Influence**. Macroscopic and microscopic characterization for two different noise amplitudes, (namely, $\Delta = 0.05$ and 0.08) of two states that for $\Delta \to 0$ and $d = 0.2$ were FS-PS (first and third columns) and chaotic (second and fourth columns). In the rows the same variables as in Fig. 4 are displayed. Macroscopic attractors displayed by reporting The variables corresponding to population 0 (resp. 1) are shown in black (resp. red). As regards the parameter values, ($g_c = 0.04$, $g_s = 0.1$) for first and third columns, and ($g_c = 0.08$, $g_s = 0.2$) for second and fourth columns. The employed noise amplitude are reported above the corresponding columns. In all cases the dilution was fixed to $d = 0.2$ and $N = 400$

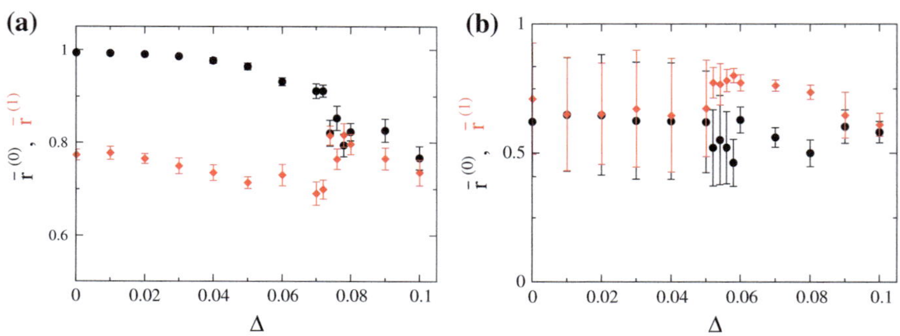

Fig. 7 In panels **a**, **b** are reported the average values of the order parameters $\bar{r}^{(0)}$ and $\bar{r}^{(1)}$ as a function of the noise Δ. As regards the parameter values, ($g_c = 0.04$, $g_s = 0.1$) in **a** and ($g_c = 0.08$, $g_s = 0.2$) in **b**. In both cases the dilution is $d = 0.2$

amplitude) the symmetry among the dynamics of the two populations. Thus in this case, the disorder can promote the emergence of a chimera-like state (a PS1-PS2) in a range of parameters where the dynamics was fully symmetric in the globally coupled deterministic set-up. For large dilution/noise the system always ends up in a partially synchronized regime. This can be explained by the fact that the stable state, for the chosen parameter and for identical coupling among neurons of both populations (namely, $g_s = g_c$), is the regime PS. Indeed, for large dilution or noise the heterogeneity in the synaptic coupling among neurons lying in one population or in another become less pronounced and the PS emerge. Therefore, disorder has at some intermediate level a constructive effect inducing the birth of a more complex (broken symmetry) state from a fully symmetric one, similarly to what reported in [19], where *coherence-resonance chimeras* have been observed.

Another interesting aspect, is that the chimera-like states PS1-PS2 are quite robust to dilution, they can be observed up to 80% of randomly broken links within each population, while previous results on phase oscillators pointed out that chimeras are observable up to 8% of dilution [6]. The origin of this noticeably difference is probably due to the fact that in this model PS states can be observed even in sparse networks with an extremely small in-degree ($K \simeq 10$) as shown in [9]. Furthermore, another stabilizing factor is the choice of the cross synaptic current, in our model the effect of one population on the neurons of the other population is mimicked via a macroscopic mean field, representing the average synaptic current.

The reported results represent only a first step in the study of the emergence of chimera states in neural systems characterized by a sparse topology and by the presence of noise. Further analysis will be required to investigate more realistic models and to understand if chimera states can have a role in the encoding of information at a population level in brain circuits.

Acknowledgements We thank E.A. Martens, A. Politi and S. Gupta for extremely useful interactions. This article is part of the research activity of the Advanced Study Group 2016 *From Microscopic to Collective Dynamics in Neural Circuits* performed at Max Planck Institute for the Physics of Complex Systems in Dresden (Germany).

Appendix: Accurate Event Driven Map for the Two Fully Coupled Populations

In the two symmetrically fully coupled populations setup here discussed we find various kind of symmetric and symmetry broken states; in particular we find synchronized states. The integration of such states can be a difficult issue, due to numerical round-off it can become extremely difficult to determine the next firing neuron, thus to increase the numerical accuracy and to avoid spurious clustering due to numerical round-off, we implemented the following integration scheme.

In particular, instead of integrating the membrane potentials, we performed the integration of the logarithm of the difference of two successive neurons. This transformation is uniquely defined in globally coupled systems, since the order of the neurons passing threshold is preserved in time. Therefore, it is possible to define an ordered list of the potentials and, on this basis, to define uniquely the "neighbours" of a neuron. Given a set of N membrane potentials $\left\{x_j^{(k)}\right\}_{j=1,\ldots,N}$, with $k = 0, 1$ depending on the considered family, we introduce at a generic time t the following $N + 1$ auxiliary variables:

$$\begin{aligned}
\omega_1^{(k)}(t) &= \ln\left[1 - x_1^{(k)}(t)\right] \\
\omega_j^{(k)}(t) &= \ln\left[x_{j-1}^{(k)}(t) - x_j^{(k)}(t)\right] \qquad j = 2, \ldots, N \\
\omega_{N+1}^{(k)}(t) &= \ln\left[x_N^{(k)}(t)\right]
\end{aligned}$$

where the threshold (resp. reset) value is $x_{th} = 1$ (resp. $x_R = 0$) and $x_1^{(k)}$ is the next to threshold neuron.

Since we would like to define an event driven map for the two coupled families, it is necessary to find which neuron is going to fire next and then evolve the membrane potentials of the two populations until the successive spike emission. The evolution of the two populations is different and it depends on the fact that the firing neuron belongs to the considered family or not. Let us schematize the algorithm in three steps:

1. As a first step we compare $x_1^{(0)}$ with $x_1^{(1)}$ to identify to which family the firing neuron belongs.
2. As a second step we check if the firing neuron belongs to a family which has already fired at the previous event or not. Depending on this, we have two possible alternatives: if the next and previous firing neurons belong to the same family we iterate the network as in point (a) below, otherwise as in point (b).
    a. Let us suppose that the firing population is the family (0). We evolve all the $N + 1$ variables $\left\{\omega_j^{(0)}\right\}_{j=1,\ldots,N+1}$ for the firing family, while, for the other family, it is sufficient to evolve just the variables $\omega_1^{(1)}$ and $\omega_{N+1}^{(1)}$. All the above mentioned variables are evolved for a lapse of time corresponding to the interval elapsed from the last firing event of the family (0).
    b. If the firing family is (1) and previously fired a neuron of family (0), the evolution is more complicated. The variables $\omega_1^{(k)}$ and $\omega_{N+1}^{(k)}$ of both families are integrated for the time interval elapsed from the last firing time of family (0). The $N - 1$ variables $\left\{\omega_j^{(1)}\right\}_{j=2,\ldots,N}$ should be instead evolved for a longer time corresponding to the last interspike interval associated to family (1), because these variables have not been updated since the last firing of family (1).

3. The firing family is iterated in the comoving frame: this amounts to update the membrane potentials and to shift the index of all neurons by one unit. The membrane potentials of the other family are updated in the fixed reference frame.
4. The simulation is iterated by repeating the above three steps.

In order to evolve the linearized system the previous algorithm is no more effective since now it is necessary to evolve all the variables at each time step in order to calculate the linearized equations in the tangent space. In this case we use directly the difference of the membrane potentials of two successive neurons instead of the logarithm. We still search for the first to fire neuron between the two populations and we treat differently the variables of the two populations at each time step depending on which neuron has emitted a spike previously. As in the previous case we employ different reference frames for the firing or not firing family (see point 3 above).

References

1. Abrams, D.M., Mirollo, R.E., Strogatz, S.H., Wiley, D.A.: Phys. Rev. Lett. **101**, 084103 (2008)
2. Angulo-Garcia, D., Torcini, A.: Phys. Rev. E **91**, 022928 (2015)
3. Aurell, E., Boffetta, G., Crisanti, A., Paladin, G., Vulpiani, A.: Phys. Rev. Lett. **77**, 1262 (1996)
4. Bolotov, M.I., Osipov, G.V., Pikovsky, A.: Phys. Rev. E **93**, 032202 (2016)
5. Bordyugov, G., Pikovsky, A., Rosenblum, M.: Phys. Rev. E **82**, 035205 (2010)
6. Laing, C.R., Rajendran, K., Kevrekidis, I.G.: Chaos **22**, 013132 (2012)
7. Laing, C.R.: Chaos **22**, 043104 (2012)
8. Loos, S.A., Claussen, J.C., Schöll, E., Zakharova, A.: Phys. Rev. E **93**, 012209 (2016)
9. Luccioli, S., Olmi, S., Politi, A., Torcini, A.: Phys. Rev. Lett. **109**, 138103 (2012)
10. Nakagawa, N., Kuramoto, Y.: Physica **80D**, 307 (1995)
11. Olmi, S., Livi, R., Politi, A., Torcini, A.: Phys. Rev. E **81**, 046119 (2010)
12. Olmi, S., Politi, A., Torcini, A.: EPL (Europhys. Lett.) **92**(6), 60007 (2010)
13. Olmi, S.: Chaos **25**, 123125 (2015)
14. Panaggio, M.K., Abrams, D.M.: Nonlinearity **28**, R67 (2015)
15. Pazó, D., Montbrió, E.: Phys. Rev. X **4**, 011009 (2014)
16. Pikovsky, A., Rosenblum, M.: Phys. Rev. Lett. **101**, 264103 (2008)
17. Rothkegel, A., Lehnertz, K.: New J. Phys. **16**, 055006 (2014)
18. Sakaguchi, H.: Phys. Rev. E **73**, 031907 (2006)
19. Semenova, N., Zakharova, A., Anishchenko, V., Schöll, E.: arXiv:1512.07036 (2015)
20. Tsigkri-DeSmedt, N.D., Hizanidis, J., Hövel, P., Provata, A.: Procedia Comput. Sci. **66**, 13–22 (2015)
21. van Vreeswijk, C.: Phys. Rev. E **54**, 5522 (1996)
22. Winfree, A.T.: The Geometry of Biological Time. Springer, Berlin (1980)
23. Zillmer, R., Livi, R., Politi, A., Torcini, A.: Phys. Rev. E **76**, 046102 (2007)

Nanotechnologies for Neurosciences

A. Aloisi, D. Pisignano and R. Rinaldi

Abstract The applications of nanotechnology in the field of neuroscience can be divided into two main strands: (i) applications in the field of basic research and (ii) applications in the clinical field. In the first area we deal with: (a) developing and applying nano-engineered materials to promote adhesion, growth and neuronal differentiation and to understand the neurobiological mechanisms underlying these processes; (b) fabricating nano-systems (for example, "nano-electrodes" implantable) for direct iteration, recording and stimulation of the neurons at the molecular level; (c) applying nano-structures and nanoscale resolution microscopy for advanced and better resolution imaging and diagnostics. In the clinical context, however, the primary goal is to limit or reverse the neurodegenerative processes. In this Lecture Note we present three different approaches at the crossing between basic research and application in clinical field. First, we report on the study of the effect of endogenous dipeptides in neurodegenerative diseases. Then we discuss some recent results in the field of the development of nano-engineered biocompatible materials ("scaffolds") that might facilitate and accelerate neuronal growth, which represents one of the fundamental objectives of modern tissue engineering. As well, we describe the synthesis of biocompatible micro- and nano-systems that can transport small molecules, drugs, immune system or stem cells, through different routes of administration, a primary goal for the treatment of a wide family of neurological disorders, as well as brain tumors. Finally, we discuss the packaging of stimuli responsive composite systems for cell and cell surrounding environment monitoring, a new road now starting to be strongly pursued.

A. Aloisi · R. Rinaldi (✉)
Institute for Microelectronics and Microsystems, CNR, Lecce, Italy
e-mail: ross.rinaldi@unisalento.it

D. Pisignano
NEST, Institute of Nanoscience, CNR, Pisa, Italy

D. Pisignano
Department of Physics, University of Pisa, Pisa, Italy

R. Rinaldi
ISUFI, University of Salento, Lecce, Italy

R. Rinaldi
Department of Mathematics and Physics "E De Giorgi", University of Salento, Lecce, Italy

F. Corinto and A. Torcini (eds.), *Nonlinear Dynamics in Computational Neuroscience*, PoliTO Springer Series, https://doi.org/10.1007/978-3-319-71048-8_6

1 Introduction

A dynamical system consists of a set of variables that describes its state and a law that describes the evolution of the state variables with time [29]. Experimental and theoretical–computational approaches in neuroscience aim is to expand the understanding of neuronal dynamics and function, in order to translate these advances into technological applications and into the diagnosis and treatment of neurologic disorders. New materials and nanoscience–based approaches involving functional peptides and biopolymers allow researchers to interface with living tissues and neuronal cells in an effective way, identifying and monitoring over time patterns of changes/interactions. Integration is possible across multiple levels (molecule–cell–tissue) of a physiological system or process and multiple levels (single element-2D assembly-3D assembly) of multi–material component/device.

In the field of basic research. Use of nano–structures and nanoscale resolution microscopy for advanced and better resolution imaging and sensing. In the clinical field. Development and application of nano–engineered materials to promote adhesion, growth and neuronal differentiation and to understand the neurobiological mechanisms underlying these processes; fabrication of nano-systems (like for example implantable "nano-electrodes") for direct iteration, recording and stimulation of the neurons at the molecular level and/or delivery of cells/drugs across the physiological barriers.

Basic research, prospective application in clinical field: topics of current research.

- The effect of dipeptides in neurodegenerative diseases.
- Micro- and nano-architectural constructs for application in neural regeneration.
- Novel Soft Materials for Neural Interfaces.

2 The Effect of Dipeptides in Neurodegenerative Diseases

This section looks at the impact of carnosine on amyloid-β peptide (Aβ) dynamics in the context of Alzheimer's disease (AD). Currently, the action of small molecules capable to hinder the structural organization of soluble or aggregating Aβ peptide during fibrillogenesis is under investigation. Atomic force microscopy, fluorescence spectroscopy and in silico molecular docking approach are exploited to calculate conformational carnosine/Aβ1–42 interactions and to show differences in the aggregation state of Aβ1–42 small pre-fibrillar structures (monomers and small oligomers) in the presence of the endogenous dipeptide beta-alanyl-L-histidine. Remarkably, carnosine has been shown to affect in vitro fibril elongation, and in silico simulations support the evidence that carnosine specifically binds to the Aβ1–42 peptide.

2.1 Amyloid Proteins and Alzheimers Disease

A number of pathologies, comprising AD, are associated with amyloid fibrils. AD is a degenerative brain ailment that impairs mainly the memory and cognitive function in the ageing. Extracellular β-amyloid deposition is one of the pathological events thought to cause neuronal damage. In their so-called amyloid state, several proteins form extended fibers, with spines consisting of many-stranded β-sheets [19]. Details of the packing interactions, by which the constituent β-strands are assembled hierarchically into protofilaments, filaments, and mature fibrils, have been given by combining structural panels from a series of experimental techniques spanning five orders of magnitude in length scale [22]. Notably, proteins amyloid state results when a segment exposes its backbone amide NH groups and C=O groups, allowing them to associate into hydrogen bonds with other protein chains. The β-sheet plane and the backbone hydrogen bonds that connect the β-strands are oriented parallel to the main fibril axis. When the nucleus is formed within a slow nucleation phase, a fast growth phase follows: single molecules can contact the nascent fibril one at a time and link at the ends of fibrils. See for a full review [19, 21]. Decreasing the generation of the amyloidogenic arrangement of proteins or increasing the clearance degree of misfolded/aggregated proteins, as well as direct inhibition of the self-assembly process, have been suggested as remedial strategy. In this perspective, constraining Aβ aggregation at the level of the monomer appears a smart approach: actually, it represents the earliest opportunity to interfere with the aggregation cascade [33, 48].

2.2 Carnosine and β–Amyloid Peptide: Interaction Dynamics

The naturally occurring carnosine is becoming a clinically accepted nutritional supplement with application over a huge range of chronic diseases. The relatively new concept of pharmacological chaperones in the treatment of certain chronic disabling diseases may hold the biological and therapeutic activities of the family of imidazole-containing peptidomimetics carcinine (β–alanylhistamine), N-acetylcarnosine (N-acetyl-β-alanylhistidine) and carnosine (β-alanyl-l-histidine) that have been described in a recent review [7]. Given the consistent body of evidence supporting the neuroprotective activity of carnosine, independent studies have recently indicated that millimolar concentrations of this dipeptide are able to effectively hinder Aβ aggregation in tube tests. Carnosine has been shown to limit the assembly of β–amyloid in the central nervous system of transgenic mice model for Alzheimers disease [11] and reduce the in vitro fibrillogenesis of β-amyloid peptide Aβ42 [3, 6]. This aptitude could be of particular effect since the direct interference with Aβ aggregation is a major target in AD therapeutics [25]. Consistent effects of carnosine are broadly shown in the literature, however the biomolecular pathways initiating these effects are still unclear. Zinc–ion modulation, upregulation of heat-shock protein expression, and/ or amplified proteolysis of the abnormal polypeptide could be at the basis of the mechanism. See, for a recent discussion authors' work in [40]. The salt bridge between the side chains of D23 and K28 is known to be involved in Aβ toxicity, as those residues are crucial in the intermolecular interactions between two adjacent Aβ42 monomers during the elongation of protofibrils. This agrees with the model of growth by single monomer addition [36] and is supported by docking analysis results which clearly show the arrangement of carnosine at this site. Specifically, the extensive and comparative molecular docking studies performed on numerous natural or synthetic recognized inhibitors of the Aβ142 fibril formation, enclose carnosine together with the top-ranked compounds. In particular, the in silico screening of Aβ142 ligands predicts this dipeptide as the best one among the natural histidine-containing tested dipeptides, and as a very good ligand when matched with other compounds capable of inhibiting the aggregation of the Aβ1–42 amyloid peptide. Interactions between the imidazole ring of the l-histidine of carnosine and the D23 residue of Aβ142, and between the β-alanine of the dipeptide and the K28 residue of Aβ142 are evidenced, supporting the potential interfering role of carnosine on fibrillogenesis. Actually, carnosine is able to disturb the growing fibril by impeding direct binding of the D23 of one Aβ142 monomer and the K28 of the contiguous monomer (Fig. 1) [3]. The evidence of a reduced Aβ142 polymerization process has been quantitatively confirmed by ThT fluorescence assay in the presence of growing concentrations of carnosine. Correspondingly, by AFM imaging of dehydrated samples, it has been observed that, in the presence of carnosine, extended Aβ1–42 fibrils items are drastically underrepresented, while few deposited aggregates predominantly include shorter fibrils and small globular formations (Fig. 1) [3].

3 Micro- and Nano-architectural Constructs for Application in Neural Regeneration

Considerable development of supramolecular biomaterials over the past decade has established, in principle, their broad utility for the design of highly controllable and highly functional composite for biomedical applications. In the future years, an increasing number of strategies based on supramolecular assembly will advance experimental devices into clinical practice, for the treatment of disease [54]. Nowadays, numerous approaches have been explored to build three-dimensional (3D) porous scaffolds with different kind of building-block materials. Actually, the soft ones play a unique role in tissue regeneration and repair: they should act as tunable biophysical and biochemical environments to direct cellular behavior and function thus promoting the replacement and regeneration of missing or injured tissues [42].

With respect to the nervous system and its unique features, both at central and peripheral level, the design of biomaterials promoting graft survival and integration as well as localized and sustained delivery of biologics to injury sites is actively being pursued. Applications of nano- and micro-scale structures, embedding growth factors or neuro-protective molecules, severely limit the damaging effects of free

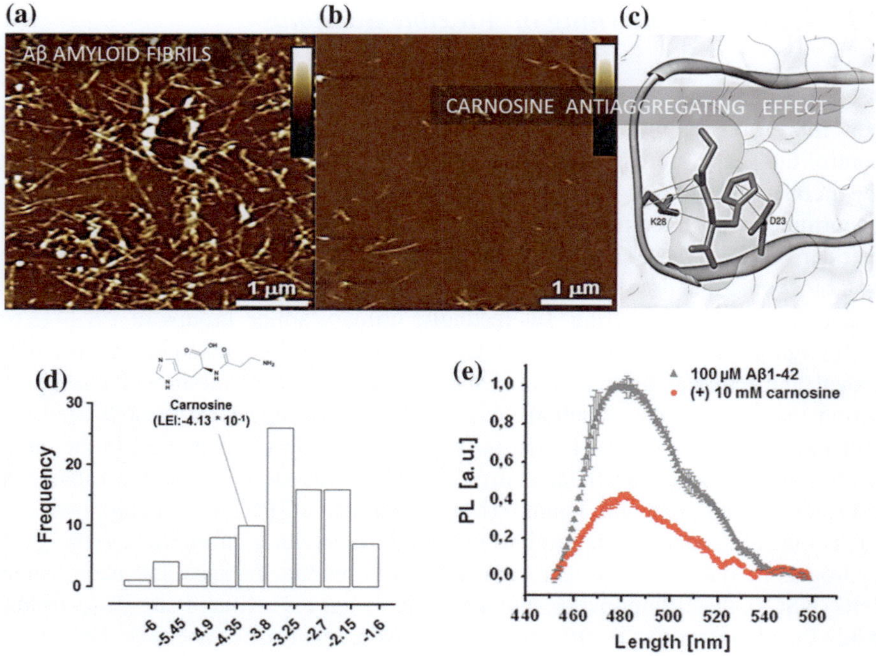

Fig. 1 Carnosine and β-amyloid peptide interaction. AFM pictures (**ab**), view of deposited Aβ1-42 amyloid aggregates, with representative fibrils from Aβ1-42 samples (A; control) and Aβ1-42 samples incubated with 10 mM carnosine (**b**). In silico analysis of carnosine/Aβ1-42 interactions. Binding models of carnosine with helical (PDB acc no.: 1Z0Q) structures of Aβ1-42 obtained by using Autodock Vina and visualized by Yasara (**c**); carnosine interacts with the folded structure in a region containing an L-aspartic acid (D23) and an L-lysine (K28). Ligand efficiency indices calculated from molecular docking analysis of the β-fibril model of Aβ1-42 (PDB: 2BEG) vs carnosine and 89 selected molecules, including carnosine-like dipeptides and natural or synthetic compounds tested for anti-amyloid aggregation effects. Ligand efficiency indices are here indicated as BE/NoC (number of carbons-based efficiency index): the more negative, the more efficient (**d**). Spectroscopic evaluation, effects of carnosine on Aβ1-42 fibrillogenesis by ThT assay (**e**). Adapted from PLoS ONE 8(7): e68159. Aloisi A, Barca A, Romano A, Guerrieri S, Storelli C, Rinaldi R, et al. (2013). doi:10.1371/journal.pone.0068159. With permission from the publisher

radicals that are generated as a result of suffering in the damaged areas [49]. In this section we will address on recent findings in biocompatible polymers as materials for scaffolds that can modify the post-injury environment, including (i) patterned flexible guide and (ii) stimuli-responsive carriers decorated 3D supports; as well (iii) drug and cell delivery vehicles.

3.1 Hydrogel Patterning on Flexible Scaffolds

A considerable body of research has recently focused on the exploitation of hydrogels for various applications in tissue engineering [51]. Moreover, the ability to control the shape, porosity, surface morphology of hydrogel scaffolds, mimicking the architecture of three-dimensional tissue at the micro/nanoscale has created new opportunities to overcome various challenges in this field. Chitosan (CS), a copolymer derived from the alkaline deacetylation of chitin, has been shown to be suitable as a biocompatible drug carrier, wound dressing material, and progressively a candidate for tissue engineering. Interestingly, primary amine groups of CS [41] or chitooligosaccharide (chitosan degradation products) [57] can support neurite outgrowth and neuronal differentiation. On the other hand, by pH variation or enzymatic action, the gelation or degradation of chitosan-based hydrogels can be modulated. CS molecular structure, similar to that of the glycosaminoglycans present in the basal lamina and in the extracellular matrix (ECM), facilitates inter alia the molecular cohesion, for example with laminin (LN) or fibronectin (FN), promoting, thanks to such molecules, adhesion, migration, and cell differentiation. In addition, varying the hydrogels modulus could direct stem cells into different lineages. For neural stem cells (NSCs), the differentiation into neuronal lineages is favoured on soft scaffolds, while the oligodendrocyte differentiation is favoured on stiffer ones [8]. Thus, fine-tuning the mechanical properties of hydrogels assumes a key role in nervous tissue regeneration. In this respect, here, we introduce a new three-dimensional construct model that allows 3D printing of CS (with peculiar morphological, mechanical and biochemical features) on a cross-linked collagen film (CL) (with different assets). Schwann cells (SCs), myelin-forming cells in the peripheral nervous system, play a fundamental role in neuron survival and function. Micro-structured CL substrates improve the response of human Schwann cells allowing them to survive and proliferate, modifying neurite outgrowth [50]. Likewise, the interaction of SCs with chitosan membranes and fibers as well as with micropatterned CS was shown in vitro [34, 59]. In particular, CS is extremely hydrophilic and it can be exploited both for its ability to expand in the presence of water and for its poly-porous matrix, in order to constitute an extracellular artificial construct suitable for neuronal growth [26]. The assembly process of molecules and biopolymers, geometrically defined at the micro-scale, traditionally involves the use of basic methods in micro- and nanotechnology (common examples are the replica moulding technique (REM), microcontact printing (μCP), microtransfer molding (μTM)) and micromolding in capillaries (MIMIC). By means of a versatile μCP modified procedure -here named Hydro-reverse μCP (Hr-μCP)- we are able to print a geometrically controlled 3D pattern of CS hydrogel on a flexible collagen film [4] (Fig. 2). A mixture of fibronectin-mimetic protein covalently linked to the chitosan, applied according to the optimized procedure of Hr-μCP, gives rise to a stable nonetheless biodegradable construct. RSC96 schwann cells, normally well cultured on collagen film, incubated with the 3D printed-film, over 3 days, show to prefer the chitosan fibronectin (CS-FN) paths during adhesion and proliferation route. Notably, SCs assume a critical role for neurite extension while FN has a key

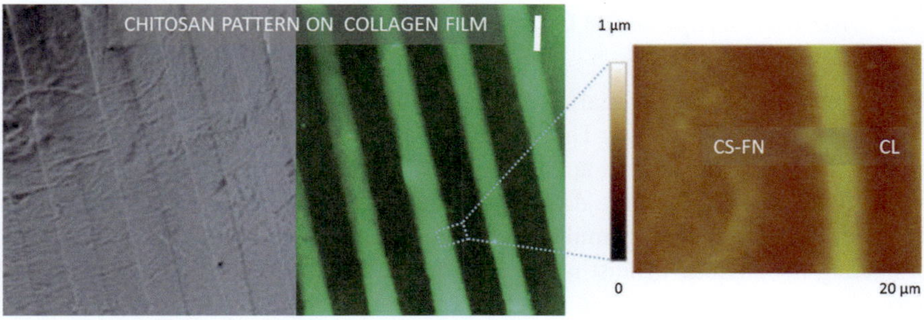

Fig. 2 Hr-μCP construct, 3D chitosan pattern on collagen film. Bright field and fluorescence microscopy of labeled chitosan. Scale bar: 100 µm. Particular of chitosan/collagen contiguous region by AFM recording in air, in tapping mode (height channel). The hydration at the interface, between the two elements of the printing (reticulated collagen film – in contact with – chitosan mixture coated PDMS replica), allows removing the collagen film with minimal resistance; the resulting stripes appear continuous (without substantial breaks) and regular in term of height profile

role in glial cells guidance, process that is favoured in the presence of topographical nanostructuring [37]. Further studies are being devoted to screen the Hr-μCP protocol performance and cell response, in function of planned strips dimension (both height and width) and matrix porosity, strictly correlated with biopolymers concentration/reticulation, pH and, especially, hydration parameters process. Why hydration is so relevant? Essentially, the particular organization of water in the hydrogel, which determines the properties of the water component, is typically different than those of bulk water; specific polymer-water interactions and the geometrical confinement of water molecules in the pores (mainly related to sorption/diffusion phenomena) influence the assets of the polymer matrix, which are different for the diverse states of water [39]. Briefly, the whole protocol has been optimized at a controlled temperature (25 °C) with the specific aim of patterning an engineered hydrogel on a reticulated collagen film, preventing any degradative process. It is significant to mention at least four critical steps of this novel method: (i) the spin-coating parameters of the polymer film on to the PDMS replica; (ii) the pre-printing collagen film hydration time, (iii) the printed collagen film detachment and (iv) the post-printing phase, crucial for the patterned structure stabilization. Finally, we should highlight the fact that by easily rolling of the ultimate assembled construct, a smart nerve conduit is achievable. It grants a microstructured parallel banding with micro or nano-porosity, along the major axis, suitable as physical cell directional guide and as responsive tool-box since functional molecules can be embedded and released, over time, in function of polymer reticulation and specific environment changes.

3.2 Stimuli–Responsive Carriers Decorated 3D Scaffolds

Three-dimensional porous scaffolds based on ECM are promising devices for tissue engineering applications. The addition of stimuli-responsive carriers (nano-and microparticles) in the current approaches to tissue reconstruction and repair brings about novel challenges in the design and conception of carrier-integrated polymer scaffolds. Notably, single or multiple biomolecules particle-based carriers allow for protecting the payload against unwanted degradation while being able to release it upon diverse stimuli, such as a pH change, protease degradation, light irradiation, magnetic field, or ultrasound [23, 28, 32, 38].

3.2.1 Layer–by–Layer Assembly

Polyelectrolyte multilayer microcapsules are assembled via the layer-by-layer (LbL) approach, alternating layers of negatively charged dextran sodium salt (DEX), or BSA FITC and positively charged Poly-L-arginine hydrochloride (PARG) on to the dextran-loaded CaCO3 cores, until two bilayers of the polyelectrolytes had been established. The sacrificial CaCO3 cores is then removed by complexation with EDTA buffer and successive washings. To obtain cationic microcapsules, an additional layer of PARG is added following core removal. By tuning the charge of the microcapsules and by varying the fabrication conditions, collagen scaffolds decorated with high or low number of microcapsules can be obtained. Cationic microcapsules bind better than that of anionic ones, and application of vacuum during scaffolds functionalization significantly hinders the attachment of the microcapsules to the collagen matrix. Remarkably, the microcapsules-integrated scaffolds fully retain their original physical properties [15].

3.2.2 Diffusional Tests

The collagen scaffolds could stably retain the microcapsules over long incubation time in Tris-HCl buffer at 37C without undergoing morphological changes, thus confirming their suitability for tissue engineering applications. These findings suggest that the combined use of scaffolds and biodegradable protein-loaded microcapsules is an encouraging strategy for the design of multifunctional microcapsules-integrated scaffolds for controlled delivery of bioactive molecules [15]. Biocompatibility of the produced composite scaffolds as well as their in vivo performances in nervous tissue regeneration studies are on-going (Fig. 3).

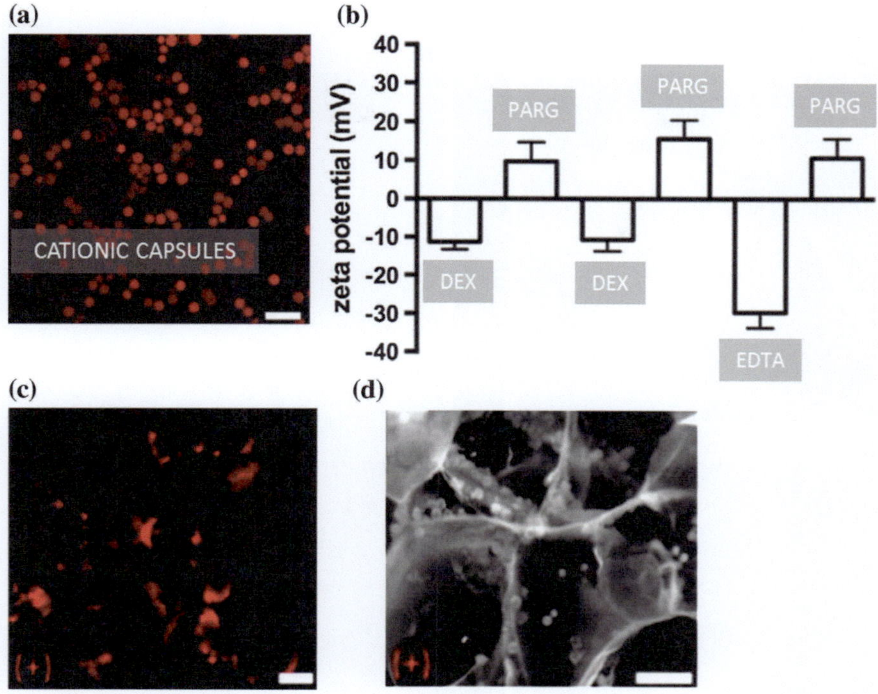

Fig. 3 Microcapsules-integrated 3D scaffolds. Fluorescence microscopy images of cationic capsules, for visualization purposes, RITC-labeled dextran conjugates were entrapped into the cavities (**a**). Zeta potential values of cationic microcapsules before and after the core removal step by EDTA (**b**). Scale bars: 10 μ m. Morphological observations by CLSM of freeze-dried 3D collagen scaffolds integrated with cationic capsules, following incubation on the plate shaker for 4h (**c**). Corresponding SEM images of scaffolds (longitudinal sections) (**d**). Scale bar: 25 μ m. Reprinted, after adaptation from [15]. Copyright (2016), with permission from Elsevier

3.3 Drug and Cell Delivery Vehicles

Hydrogels can likewise serve as drug delivery platforms for biomolecular therapeutics into the CNS/PNS or as scaffolds for delivering cells to repair the injured tissue [35]. As natural, biocompatible, biodegradable and non-immunogenic building blocks, alginate, (anionic, linear, polymer of varying ratios of guluronic and mannuronic acid units), and chitosan (linear polysaccharide composed of randomly distributed (1-4)-linked D-glucosamine and N-acetyl-D-glucosamine) are been currently exploited. Ionotropic pre-gelation of the alginate core, followed by chitosan polyelectrolyte complexation, allow to encapsulate selected active molecules by means of physical entrapment and electrostatic interactions within sub-micron sized hydrogel vesicles. Here we present a microfluidic-assisted assembly method of nano- and micro-vesicles (Fig. 4) -under sterile, closed environment and gas exchange adjustable conditions- a critical issue, when the cargo to be upload is very

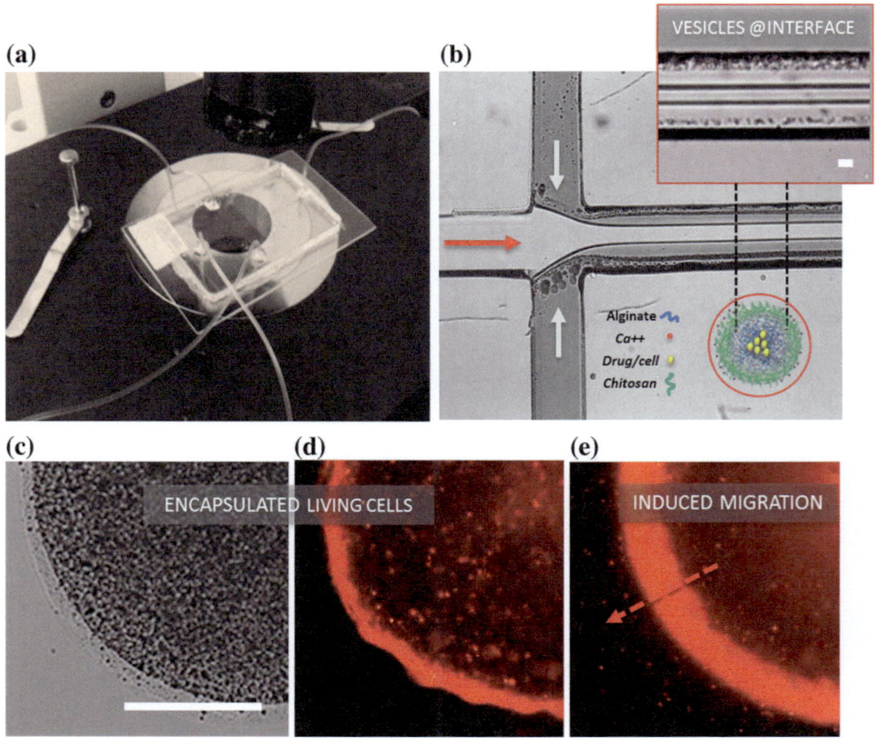

Fig. 4 Microfluidics-aided synthesis of chitosan/alginate complexes. Image of the device set up (**a**); flow dynamics, optical microscope magnification (**b**); alginate-drug or cell loaded-vesicles (graphical schema in B, circular inset) are formed at the interface of the microchannel (b squared inset, scale bar 20 μm). Microfluidic mixing grants a good control of flow rate and reaction time; low reagent amount; control of vesicles size from hundred nanometers to hundred micrometers. Post-synthesis, oppositely charged polymer LbL coating approach allows modulation of cargo release in function of polymer reticulation and chemical properties. Schema, at the macro-scale: encapsulated living cells (**c-e**) are able to migrate across hydrogels micrometric mesh under chemiotactic stimuli (**e**); modulation of polymer reticulation represents a key topic. Scale bar: 500 μm [5]

sensitive [5]. Microfluidic dynamics, due to the dimensions involved, are mainly characterized by laminar flow, which results in unique fluid morphologies and the production of polymer structures with an improved control over their sizes, size distributions, and compositions (by adjusting flow velocity as well biopolymer and reticulating agent concentrations). Furthermore, as active molecules cargo amount mainly influences the final -drug delivery device- product costs, the purpose of the novel system is to obtain a very efficient drug encapsulation in order to improve the efficiency of encapsulated active drug/vesicle volume ratio, even starting from low concentrations [45]. Polymer/polymer and polymer/drug mass ratio relationship are crucial in order to attain the optimum in terms of shuttle size and cargo concentration. Modulating polymer reticulation conditions, it become possible to control drug loading efficiency as well as drug delivery dynamics. Given the undoubtedly favourable properties of alginate and chitosan, different regulatory issues should turn into a relevant starting point for discussion among worldwide regulatory bodies, drug policymakers, and biopharmaceutical companies in pursuing suitable biopharmaceutical products based on biopolymer like chitosan, alginate and their derivatives. On a parallel route, cellular therapy has emerged as a treatment choice for repair and regeneration of nerve injuries. However, transplantation of stem cells to the injured sites showed poor cell survival and engraftment [56]. In this regard, the arrangement of functional biomaterials and cell delivery is a promising strategy for CNS repair. The ECM is a dynamic environment that continuously provides physical and chemical cues to embedded cells. In order to mimic the dynamic nature of the ECM, many reversible chemistries have been incorporated into hydrogels to control cell spreading related to biochemical ligand presentation and matrix mechanics [44]. Culture medium nutrients-enriched alginate efficiently works in terms of living cells embedding. Exploiting fluidic device, cells are easily encapsulated (Fig. 4) into biocompatible platforms suitable as mechanical and feeding support. Single or more encapsulated astroglial cells vitality is comparable with that of standard in vitro culture. Additional chitosan coating still permits cells to survive and grow up; moreover it offer a physical hedge -fine adjustable- in terms of elasticity and porosity, able to mimic physiologic barriers across which cells can move under specific stimuli [5].

What is extremely attractive is that these type of platforms are tailorable, for multifaceted requests. Some possible applications include: (1) encapsulated cell delivery for tissue engineering, recapitulating and improving the condition required for neuronal cells survival; (2) in vitro 3D reproduction of BBB barrier (at the interface of endothelial-glial/neuronal cells) for in vitro screening of pharmaceuticals; (3) single cell encapsulation for high sensitive living-sensor toolbox.

4 Novel Soft Materials for Neural Interface

This section mentions recent progress in fabricating micro and nano-systems as per two-way communication between electronic devices and biological tissues for direct iteration, recording and stimulation of the neurons, as well as in fabricating analyte sensors and stimuli-responsive drug-loaded structures with conceivable application in nervous system.

Both compliant mechanical structures and smart soft materials will be presented, as recent strategy suitable for spatiotemporal resolution across all relevant scales -at the interface- with biological systems.

4.1 *Stretchable Electronics*

Systems that offer elastic, low modulus responses to large strain deformations are required in order to conform to cells and biological tissues while keeping high performance in their electronic functionality. Multi-faceted and circular relationships between mechanics and biology are crucial in the process structure function paradigm of engineering [18]. For instance, mechanical interaction concerning the graft and the surrounding tissues might involve the weight, shape and flexibility of the material components. Cell and tissue damage (as a result of the implantation event initiating the inflammatory cascade) have been assigned to mechanical gap leading to structural mismatching between implanted materials and devices and the targeted biological environment. The device architectures and the material selections must be considered together for minimally invasive and mechanically stable technology. In particular, this would lead to achieving soft neural interfaces tailorable modulus, where inorganic active materials can be combined with low modulus polymers [30]. Up-and-coming approaches to this aim exploit advanced composites which can be attained by different methods, including dispersing nanomaterials (nanowires/nanotubes/graphene) into elastomeric matrices [2, 43, 46, 47, 55] and designing thin layers of electronic materials which can be lithographically defined into two dimensional (2D) -fractal-based- filamentary mesh layouts [9, 20, 24, 52].

4.2 *Soft Multi–analyte Sensing Platform*

Among micro- and nano-technological approaches for neuroscience, optical strategies are also exploitable for both recording and surveying changes in neural activity and cell surrounding environment. Particularly, packaging of sensor dyes into nano- and micro-structures display fine fluorescence changes (quenching/enhancement upon binding of the ligand) that are correlated to the presence/level fluctuations of the specific molecule, and these tools can be used and optimized as suitable optical reporter.

4.2.1 Capsule-Based Fluorescent Ratiometric Sensors

Stimuli responsive, fluorescent probes have attracted extensive scientific interest in recent years, due to their potential use as building blocks within sensing device applications. This is especially relevant in view of the development of multi analyte-responsive systems. In this framework, a very interesting class of materials which can be engineered to exhibit multiple functions are polyelectrolyte, multilayer capsules produced by Layer-by-Layer (LbL) assembly [12]. Selected probes can be loaded during different steps of the LbL procedure, namely before co-precipitation of the templates, onto pre-formed templates/between the layers of the shell by electrostatic interactions, or within hollow capsules made up by multilayer shells sensitive to pH, salt, light and temperature (swelling/shrinking processes). The specific strategy to

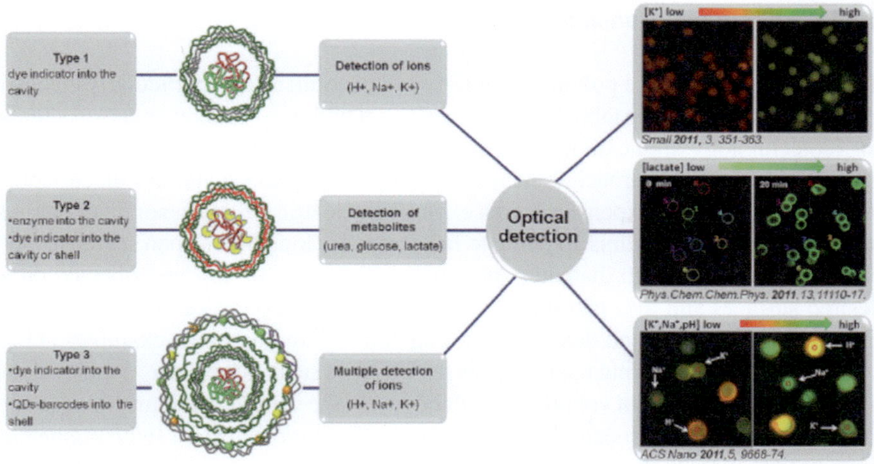

Fig. 5 Capsule-based sensors. Type 1, ratiometric readout is performed at single capsule level via optical recording of the fluorescence changes displayed by the sensor and is used to quantify the analyte concentration; type 2, reaction between the substrate and the enzyme is valued by fluorescence changes of the indicator dyes which are correlated to the reaction products; type 3, multiple detection and quantification of individual analytes is achieved by combining the ratiometric readouts of the cavities (sensing assay) with readouts of the shells (fluorescent barcodes). Reprinted from [14]. Copyright (2014), with permission from Elsevier

be pursued strongly depends on the physical and chemical properties of the cargo molecules. The question might be: why polyelectrolyte multilayer capsules are suitable as optical reporters? Actually, micrometer-sized polyelectrolyte capsule-based sensors offer the possibility of loading a wide range of analyte-sensitive, fluorescent probes within their cavities while maintaining their unaltered platform shell. In this way, capsules can be taken up by cells via the same uptake route, but their cores can display different sensing properties. Analyte-sensitive fluorophores can be loaded together with reference labels (whose emission is analyte-concentration insensitive) in the same sensor carrier. This allows for carrying out multiplex real-time biosensing ratiometric evaluation of cellular functions [13].

Three concepts of capsule-based sensors are presented in Fig. 5:

(i) indicator and reference dyes (e.g., FITC/RITC), or ratiometric indicator dyes (e.g., SNARF) coupled with high-molecular-mass polymers (e.g. dextran), are embedded in the cavities of the capsules.
(ii) metabolite-degrading enzymes are located inside the cavity, the indicator dye is co-loaded into the cavity or into the shell.
(iii) the sensor components (e.g., indicator dyes) are confined into the inner cavity of a triple-shell capsule, and the outermost shell is tagged with a unique quantum-dot based barcode.

See for a comprehensive review [14].

4.2.2 Electrospun Nanofibers

Technologies to produce polymer nanorods and nanofibers, and particularly electrospinning are now well-established nanofabrication routes to generate nanostructures for biotechnology applications. Rediscovered in 1995 in the form by the Reneker group [17] as novel approach to nanofabrication on polymer materials, electrospinning is versatile and inexpensive, and it exploits an electric field to generate a charged jet of polymer solution, thus allowing to fabricate very long fibers upon solvent evaporation. Nanofibers with diameters possibly well below 100 nm are collected on a desired surfaces and they can then be used for various application. Applications in tissue engineering, wound dressing and drug delivery are just some examples [1].

Remarkably, by varying the processing conditions such as flow rate, applied electric field strength, solvent volatility, conductivity of the solution, polymer concentration, the size of the produced fibers can be controlled to some extent (with diameters ranging from microns to tens of nm), and surface features are achievable, resulting in structures with high surface-to-volume ratios and porosity. These features potentially promote the transport of ions/small molecules/fluids across fibrous nets constituted by complex assemblies of nanostructures, thus making electrospun composite fibers (Fig. 6) very suitable for the development of ultrasensitive and faster reactive sensors [10, 16, 31, 53, 58].

This field is also going to merge with polymer research, where much progress is being made in the area of so-called stimuli responsive composites, namely in novel materials capable of changes (in terms of strength/color/morphology) upon exposure of micro- or nano-structures to specific external signals. The combination of these two highly interesting research topics, i.e. electrospun nanofibers and stimuli-responsive polymers, represents a new road now starting to be vigorously pursued [27].

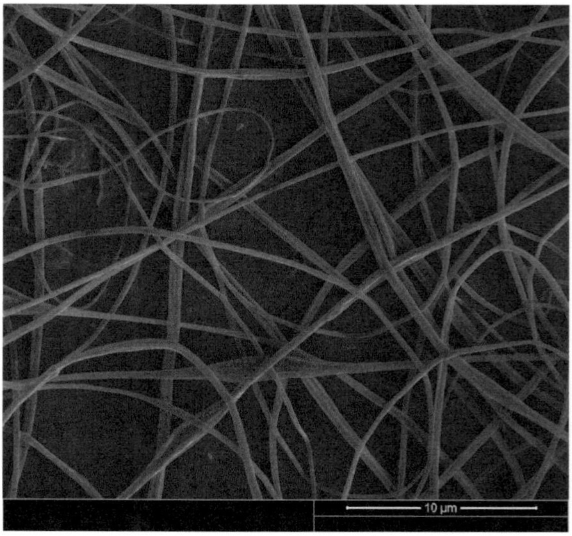

Fig. 6 Exemplary SEM micrograph of electron nanofibers made of polyvinyl pyrrolidone embedding a conductive polymer. Credit: Dr. M. Moffa

5 Discussion

The application of micro- and nano-technological approaches in cell biology and physiology allows controlled interactions at the molecular level. Here we have briefly presented examples of current research intended to better interact with biological structures: high-resolution imaging and fabrication technologies, as well smart materials and molecules - for targeted action in neuroprotection and neural regeneration.

References

1. Agarwal, S., Wendorff, J.H., Greiner, A.: Use of electrospinning technique for biomedical applications. Polymer **49**(26), 5603–5621 (2008)
2. Ahn, B.Y., Duoss, E.B., Motala, M.J., Guo, X., Park, S.-I., Xiong, Y., Yoon, J., Nuzzo, R.G., Rogers, J.A., Lewis, J.A.: Omnidirectional printing of flexible, stretchable, and spanning silver microelectrodes. Science **323**(5921), 1590–1593 (2009)
3. Aloisi, A., Barca, A., Guerrieri, S., Romano, A., Storelli, C., Rinaldi, R., Verri, T.: Anti-aggregating effect of the naturally occurring dipeptide carnosine on aβ1-42 fibril formation. PLoS One **8**(7), 68159 (2013)
4. Aloisi, A., Rizzuti, I., Madaghiele, M., Salvatore, L., Sannino, A., Rinaldi, R.: Data not published elsewhere before published (2015)
5. Aloisi, A., Toma, C. C., Di Corato, R., Rinaldi, R.: Microfluidics and BIO-encapsulation for drug- and cell-therapy. Proc. SPIE 10364, Organic Sensors and Bioelectronics X, 103640O (2017). https://doi.org/10.1117/12.2274018
6. Attanasio, F., Convertino, M., Magno, A., Caflisch, A., Corazza, A., Haridas, H., Esposito, G., Cataldo, S., Pignataro, B., Milardi, D., et al.: Carnosine inhibits aβ42 aggregation by perturbing the h-bond network in and around the central hydrophobic cluster. ChemBioChem **14**(5), 583–592 (2013)
7. Babizhayev, M., Yegorov, Y.: Advanced drug delivery of n-acetylcarnosine (n-acetyl-beta-alanyl-l-histidine), carcinine (beta-alanylhistamine) and l-carnosine (beta-alanyl-l-histidine) in targeting peptide compounds as pharmacological chaperones for use in tissue engineering, human disease management and therapy: from in vitro to the clinic. Recent Pat. Drug Deliv. Formul. **4**(3), 198–230 (2010)
8. Brännvall, K., Bergman, K., Wallenquist, U., Svahn, S., Bowden, T., Hilborn, J., Forsberg-Nilsson, K.: Enhanced neuronal differentiation in a three-dimensional collagen-hyaluronan matrix. J. Neurosci. Res. **85**(10), 2138–2146 (2007)
9. Brosteaux, D., Axisa, F., Gonzalez, M., Vanfleteren, J.: Design and fabrication of elastic interconnections for stretchable electronic circuits. IEEE Electron Device Lett. **28**(7), 552–554 (2007)
10. Cho, N.G., Woo, H.-S., Lee, J.-H., Kim, I.-D.: Thin-walled nio tubes functionalized with catalytic pt for highly selective c 2 h 5 oh sensors using electrospun fibers as a sacrificial template. Chem. Commun. **47**(40), 11300–11302 (2011)
11. Corona, C., Frazzini, V., Silvestri, E., Lattanzio, R., La Sorda, R., Piantelli, M., Canzoniero, L.M., Ciavardelli, D., Rizzarelli, E., Sensi, S.L.: Effects of dietary supplementation of carnosine on mitochondrial dysfunction, amyloid pathology, and cognitive deficits in 3xtg-ad mice. PloS one **6**(3), 17971 (2011)
12. Decher, G.: Fuzzy nanoassemblies: toward layered polymeric multicomposites. Science **277**(5330), 1232–1237 (1997)
13. del Mercato, L.L., Abbasi, A.Z., Parak, W.J.: Synthesis and characterization of ratiometric ion-sensitive polyelectrolyte capsules. Small **7**(3), 351–363 (2011)

14. del Mercato, L.L., Ferraro, M.M., Baldassarre, F., Mancarella, S., Greco, V., Rinaldi, R., Leporatti, S.: Biological applications of lbl multilayer capsules: from drug delivery to sensing. Adv. Colloid Interface Sci. **207**, 139–154 (2014)
15. del Mercato, L.L., Passione, L.G., Izzo, D., Rinaldi, R., Sannino, A., Gervaso, F.: Design and characterization of microcapsules-integrated collagen matrixes as multifunctional three-dimensional scaffolds for soft tissue engineering. J. Mech. Behav. Biomed. Mater. **62**, 209–221 (2016)
16. Ding, B., Wang, M., Yu, J., Sun, G.: Gas sensors based on electrospun nanofibers. Sensors **9**(3), 1609–1624 (2009)
17. Doshi, J., Reneker, D.H.: Electrospinning process and applications of electrospun fibers. J. Electrost. **35**(2–3), 151–160 (1995)
18. Egan, P., Schunn, C., Cagan, J., LeDuc, P.: Improving human understanding and design of complex multi-level systems with animation and parametric relationship supports. Des. Sci. **1**, (2015)
19. Eisenberg, D., Jucker, M.: The amyloid state of proteins in human diseases. Cell **148**(6), 1188–1203 (2012)
20. Fan, J.A., Yeo, W.-H., Su, Y., Hattori, Y., Lee, W., Jung, S.-Y., Zhang, Y., Liu, Z., Cheng, H., Falgout, L., et al.: Fractal design concepts for stretchable electronics. Nat. Commun. **5**, 3266 (2014)
21. Fändrich, M., Schmidt, M., Grigorieff, N.: Recent progress in understanding alzheimer's β-amyloid structures. Trends Biochem. Sci. **36**(6), 338–345 (2011)
22. Fitzpatrick, A.W., Debelouchina, G.T., Bayro, M.J., Clare, D.K., Caporini, M.A., Bajaj, V.S., Jaroniec, C.P., Wang, L., Ladizhansky, V., Müller, S.A., et al.: Atomic structure and hierarchical assembly of a cross-β amyloid fibril. Proc. Natl. Acad. Sci. **110**(14), 5468–5473 (2013)
23. Ganta, S., Devalapally, H., Shahiwala, A., Amiji, M.: A review of stimuli-responsive nanocarriers for drug and gene delivery. J. Controll. Release **126**(3), 187–204 (2008)
24. Gray, D.S., Tien, J., Chen, C.S.: High-conductivity elastomeric electronics. Adv. Mater. **16**(5), 393–397 (2004)
25. Hamley, I.W.: The amyloid beta peptide: a chemists perspective. role in alzheimers and fibrillization. Chem. Rev. **112**(10), 5147–5192 (2012)
26. Hsieh, F.-Y., Tseng, T.-C., Hsu, S.-h: Self-healing hydrogel for tissue repair in the central nervous system. Neural Regen. Res. **10**(12), 1922 (2015)
27. Huang, C., Soenen, S.J., Rejman, J., Lucas, B., Braeckmans, K., Demeester, J., De Smedt, S.C.: Stimuli-responsive electrospun fibers and their applications. Chem. Soc. Rev. **40**(5), 2417–2434 (2011)
28. Iafisco, M., Sandri, M., Panseri, S., Delgado-Lpez, J.M., Gmez-Morales, J., Tampieri, A.: Magnetic bioactive and biodegradable hollow fe-doped hydroxyapatite coated poly (l-lactic) acid micro-nanospheres. Chem. Mater. **25**(13), 2610–2617 (2013)
29. Izhikevich, E.M.: Dynamical Systems in Neuroscience. MIT press, London (2007)
30. Jeong, J.-W., Shin, G., Park, S.I., Yu, K.J., Xu, L., Rogers, J.A.: Soft materials in neuroengineering for hard problems in neuroscience. Neuron **86**(1), 175–186 (2015)
31. Kim, I.-D., Rothschild, A.: Nanostructured metal oxide gas sensors prepared by electrospinning. Polym. Adv. Technol. **22**(3), 318–325 (2011)
32. Lam, J., Lu, S., Kasper, F.K., Mikos, A.G.: Strategies for controlled delivery of biologics for cartilage repair. Adv. Drug Deliv. Rev. **84**, 123–134 (2015)
33. Lemkul, J.A., Bevan, D.R.: The role of molecular simulations in the development of inhibitors of amyloid β-peptide aggregation for the treatment of alzheimers disease. ACS Chem. Neurosci. **3**(11), 845–856 (2012)
34. Li, G., Zhao, X., Zhao, W., Zhang, L., Wang, C., Jiang, M., Gu, X., Yang, Y.: Porous chitosan scaffolds with surface micropatterning and inner porosity and their effects on schwann cells. Biomaterials **35**(30), 8503–8513 (2014)
35. Li, Y., Rodrigues, J., Tomas, H.: Injectable and biodegradable hydrogels: gelation, biodegradation and biomedical applications. Chem. Soc. Rev. **41**(6), 2193–2221 (2012)

36. Lührs, T., Ritter, C., Adrian, M., Riek-Loher, D., Bohrmann, B., Döbeli, H., Schubert, D., Riek, R.: 3d structure of alzheimer's amyloid-β (1–42) fibrils. Proc. Natl. Acad. Sci. U. S. A. **102**(48), 17342–17347 (2005)
37. Mukhatyar, V.J., Salmerón-Sánchez, M., Rudra, S., Mukhopadaya, S., Barker, T.H., García, A.J., Bellamkonda, R.V.: Role of fibronectin in topographical guidance of neurite extension on electrospun fibers. Biomaterials **32**(16), 3958–3968 (2011)
38. Mura, S., Manconi, M., Sinico, C., Valenti, D., Fadda, A.M.: Penetration enhancer-containing vesicles (pevs) as carriers for cutaneous delivery of minoxidil. Int. J. Pharm. **380**(1–2), 72–79 (2009)
39. Pissis, P., Kyritsis, A.: Hydration studies in polymer hydrogels. J. Polym. Sci. Part B: Polym. Phys. **51**(3), 159–175 (2013)
40. Preedy, V.R.: Imidazole Dipeptides: Chemistry, Analysis. Function and Effects. Royal Society of Chemistry, Cambridge (2015)
41. Ren, Y.-J., Zhang, H., Huang, H., Wang, X.-M., Zhou, Z.-Y., Cui, F.-Z., An, Y.-H.: In vitro behavior of neural stem cells in response to different chemical functional groups. Biomaterials **30**(6), 1036–1044 (2009)
42. Rnjak-Kovacina, J., Wray, L.S., Burke, K.A., Torregrosa, T., Golinski, J.M., Huang, W., Kaplan, D.L.: Lyophilized silk sponges: a versatile biomaterial platform for soft tissue engineering. ACS Biomater. Sci. Eng. **1**(4), 260–270 (2015)
43. Robinson, A.P., Minev, I., Graz, I.M., Lacour, S.P.: Microstructured silicone substrate for printable and stretchable metallic films. Langmuir **27**(8), 4279–4284 (2011)
44. Rosales, A.M., Anseth, K.S.: The design of reversible hydrogels to capture extracellular matrix dynamics. Nat. Rev. Mater. **1**(2), 15012 (2016)
45. Sallustio, F., Curci, C., Aloisi, A., Toma, C.C, Marulli, E., Serino, G., Cox, S.N., De Palma, G., Stasi, A., Divella, C., Rinaldi, R., Schena, F.P.: Inhibin-A and Decorin secreted by human adult renal stem/progenitor cells through the TLR2 engagement induce renal tubular cell regeneration. Sci Rep. **7**(1), 8225 (2017). https://doi.org/10.1038/s41598-017-08474-0
46. Sekitani, T., Noguchi, Y., Hata, K., Fukushima, T., Aida, T., Someya, T.: A rubberlike stretchable active matrix using elastic conductors. Science **321**(5895), 1468–1472 (2008)
47. Sekitani, T., Nakajima, H., Maeda, H., Fukushima, T., Aida, T., Hata, K., Someya, T.: Stretchable active-matrix organic light-emitting diode display using printable elastic conductors. Nat. Mater. **8**(6), 494–499 (2009)
48. Sievers, S.A., Karanicolas, J., Chang, H.W., Zhao, A., Jiang, L., Zirafi, O., Stevens, J.T., Münch, J., Baker, D., Eisenberg, D.: Structure-based design of non-natural amino-acid inhibitors of amyloid fibril formation. Nature **475**(7354), 96 (2011)
49. Tam, R.Y., Fuehrmann, T., Mitrousis, N., Shoichet, M.S.: Regenerative therapies for central nervous system diseases: a biomaterials approach. Neuropsychopharmacology **39**(1), 169 (2014)
50. Van Neerven, S.G., Krings, L., Haastert-Talini, K., Vogt, M., Tolba, R.H., Brook, G., Pallua, N., Bozkurt, A.: Human schwann cells seeded on a novel collagen-based microstructured nerve guide survive, proliferate, and modify neurite outgrowth. BioMed Res. Int. **2014**, (2014)
51. Vlierberghe, S., Dubruel, P., Schacht, E.: Biopolymer-based hydrogels as scaffolds for tissue engineering applications: a review. Biomacromolecules **12**(5), 1387–1408 (2011)
52. Vanfleteren, J., Gonzalez, M., Bossuyt, F., Hsu, Y.-Y., Vervust, T., De Wolf, I., Jablonski, M.: Printed circuit board technology inspired stretchable circuits. MRS Bull. **37**(3), 254–260 (2012)
53. Wang, X., Drew, C., Lee, S.-H., Senecal, K.J., Kumar, J., Samuelson, L.A.: Electrospun nanofibrous membranes for highly sensitive optical sensors. Nano Lett. **2**(11), 1273–1275 (2002)
54. Webber, M.J., Appel, E.A., Meijer, E., Langer, R.: Supramolecular biomaterials. Nat. Mater. **15**(1), 13 (2016)
55. Wu, H., Kong, D., Ruan, Z., Hsu, P.-C., Wang, S., Yu, Z., Carney, T.J., Hu, L., Fan, S., Cui, Y.: A transparent electrode based on a metal nanotrough network. Nat. Nanotechnol. **8**(6), 421 (2013)
56. Wu, K.H., Mo, X.M., Han, Z.C., Zhou, B.: Stem cell engraftment and survival in the ischemic heart. Ann. Thorac. Surg. **92**(5), 1917–1925 (2011)

57. Yang, Y., Liu, M., Gu, Y., Lin, S., Ding, F., Gu, X.: Effect of chitooligosaccharide on neuronal differentiation of pc-12 cells. Cell Biol. Int. **33**(3), 352–356 (2009)
58. Yoon, J., Chae, S.K., Kim, J.-M.: Colorimetric sensors for volatile organic compounds (vocs) based on conjugated polymer-embedded electrospun fibers. J. Am. Chem. Soc. **129**(11), 3038–3039 (2007)
59. Yuan, Y., Zhang, P., Yang, Y., Wang, X., Gu, X.: The interaction of schwann cells with chitosan membranes and fibers in vitro. Biomaterials **25**(18), 4273–4278 (2004)

Memristor and Memristor Circuit Modelling Based on Methods of Nonlinear System Theory

A. Ascoli, R. Tetzlaff and M. Biey

Abstract Tremendous efforts are made towards the development and realization of memristors for memory technology. Furthermore, memristor-based neuron and synapse models are considered in several investigations on neuromorphic systems. Some work is devoted to take advantage of peculiar nonlinear dynamics emerging in memristors to extend or improve the functionalities of state-of-the-art circuits and systems, both in the digital domain, where their use has been proposed in logic gate design, and in the analog domain, where volatile memristors with negative differential resistance effect, capable to amplify infinitesimal fluctuations of energy, have been adopted to design interesting transistor-less circuits. While in an increasingly number of cases, memristor models based on charge transport phenomena can be verified through measurements to a high degree of accuracy, the systematic design of circuits exploiting the rich dynamical behavior of memristors is still restricted to certain cases. For some memristors, even circuits with a few number of elements may exhibit different nonlinear phenomena, which can be only described by results obtained in numerical simulations. In this manuscript a nonlinear system theory-based approach will be introduced and discussed in detail. It is based on the determination of nonlinear characteristic functions allowing the characterization of circuit properties to a high degree of accuracy. These real-valued functions represent circuit elements or sub-circuits, and may be used in an automated circuit design approach. Results showing the performance of the proposed method will be given and discussed in this contribution.

A. Ascoli (✉) · R. Tetzlaff
Faculty of Electronics and Electrotechnics, Technische Universität Dresden, Dresden, Germany
e-mail: alon.ascoli@tu-dresden.de

R. Tetzlaff
e-mail: ronald.tetzlaff@tu-dresden.de

M. Biey
Department of Electronics and Telecommunications, Politecnico di Torino, Turin, Italy
e-mail: mario.biey@polito.it

F. Corinto and A. Torcini (eds.), *Nonlinear Dynamics in Computational Neuroscience*, PoliTO Springer Series, https://doi.org/10.1007/978-3-319-71048-8_7

1 Introduction

After the first conscious discovery of memristor fingerprints [1] in nature [2], the interest in the application fields of resistance switching memories [3] has been growing at an exceptional rate in academics as well as in industry. The past few years have witnessed a considerable progress in the investigation of properties of nano-scale two-terminal elements with memristor behaviour, and a substantial parallel advancement in the exploration of their capabilities for various circuit applications. Despite there is still a long way [4] towards a comprehensive understanding of the physical mechanisms at the origin of the nonlinear dynamics [5] emerging in the devices, tremendous efforts are made to establish solid theoretic foundations on memristors. These studies are instrumental for the development of systematic analysis and synthesis techniques for memristor-based circuits and systems.

Standard circuit and theory analysis techniques are not suitable to characterize the nonlinear behaviour of memristor devices and circuits. Conversely, nonlinear system theoretic concepts constitute a relevant knowledge base for the description of memristor dynamics [6–9]. In this respect, the Volterra series paradigm [10] provides a solid theoretical framework for this type of modelling activity.

The Volterra theory models the output of a nonlinear electronic system as an algebraic sum of multi-dimensional convolution integrals involving input and representative functions, known as Volterra kernels and depending on circuit parameters only. It then provides an alternative mathematical description to the classical Differential Algebraic Equation (DAE)-based model stemming from standard application of Kirchhoff's laws to the circuit under modelling. The number of kernels needed to reproduce the dynamics of the circuit depends on the desired level of accuracy requested in the modelling process. The application of the Volterra series theory to nonlinear circuits provides some advantages over conventional modelling techniques, including the items listed below.

- The nonlinear electronic system under modelling is completely characterized once the set of representative Volterra kernels, depending on circuit parameters only, is identified. Its response to any excitation of interest is directly obtainable from a sum of multi-dimensional integrals, and, in some cases, may be computed without the need to run any numerical simulation. For a memristor device or circuit this is particularly advantageous, since the numerical integration of its DAE set often leads to the appearance of convergence issues.
- The theory of Volterra covers the model of all possible interconnections in system theory (including cascade, parallel, and unity-gain negative feedback couplings). Therefore the kernel-based mathematical description of large circuits made up of interconnections among small sub-circuits only requires the determination of a set of representative functions for each sub-circuit. Storing the kernel sets for distinct memristor devices and circuits in an apposite kernel library, an automated memristor circuit and system design methodology may be envisioned.
- The Volterra series paradigm may speed up the process of circuit design optimisation. Once the kernel set of a certain circuit are obtained, expressions for

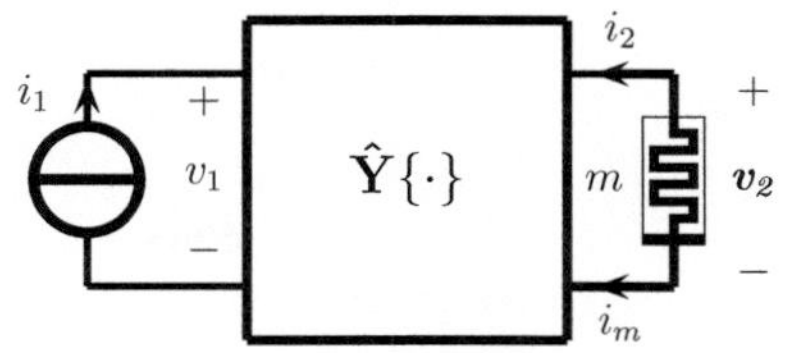

Fig. 1 Topology of the class of memristor circuits under characterization. $\hat{\mathbf{Y}}\{\cdot\}$ denotes the matrix of integro-differential operators descending from the current-voltage representation of the two-port

performance indicators, typically listed in the circuit specifications (e.g. total harmonic distortion, and signal-to-noise ratio), may be expressed as a function of the kernels, and, consequently, in terms of the circuit parameters. As a result, the circuit may be optimized by applying appropriate numerical techniques directly to the formulas of the performance indicators, rather than adopting classical iterative procedures based upon lengthy circuit simulations.

This work presents a rigorous study shedding light into the capability of the aforementioned nonlinear system theoretic paradigm to represent the dynamics of a class of memristor circuits. In the first part of these lecture notes (Sect. 2) the analysis is focused on single memristor circuits. The second part (Sect. 3) shows how the modelling technique may be extended to multi-memristor electronic systems, taking the class of two-memristor circuits as case study. Each of these two parts explain in detail the systematic procedure of the modelling approach, from the identification of relevant dynamic equations, to the determination of the circuit response to excitations of interest, explaining in depth all the mathematical steps in the derivation of analytical formulas for the system Volterra kernels. In order to highlight the pedagogical nature of this work, each of the two parts of these lecture notes is self-contained.

2 One-Memristor Circuits

The first part of this paper describes a procedure for the determination of the Volterra kernels of the class of circuits falling into the topology of Fig. 1, which consists of a current-driven linear and dynamic two-port loaded by a memristor.

The next section introduces the model adopted for the memristor.

2.1 Memristor Model

The memristor studied in this work falls into the class of *ideal generic memristors* [11]. The DAE set describing its nonlinear dynamics may be cast as

$$\frac{dw}{dt} = f(w)i_m, \tag{1}$$

$$\tilde{v}_m = \tilde{M}(w)i_m, \tag{2}$$

where w, $\tilde{v}_m$, and i_m respectively denote memristor state, voltage, and current (the variable controlling the device operation), while $f(w)$ and $\tilde{M}(w)$, are the state evolution and memristance functions [12]. As an example, the latter functions may be chosen as

$$f(w) = 3w^{\frac{2}{3}}, \tag{3}$$

$$\tilde{M}(w) = m_0 + m_1 w^{\frac{1}{3}} + m_2 w^{\frac{2}{3}} + m_3 w^{\frac{3}{3}}, \tag{4}$$

where coefficients m_0, m_1, m_2, and m_3 assume values in $\mathbb{R}_{0,+}$. It may be proved that in this case the memristor charge q_m is related to the state w through $q_m = w^{\frac{1}{3}}$. It follows that the memristance in Eq. (4) as well as the memristor flux $\tilde{\varphi}_m$ may be expressed in terms of charge as follows:

$$\tilde{M}(q_m) = m_0 + m_1 q_m + m_2 q_m^2 + m_3 q_m^3, \tag{5}$$

$$\tilde{\varphi}_m(q_m) = m_0 q_m + \frac{m_1}{2} q_m^2 + \frac{m_2}{3} q_m^3 + \frac{m_3}{4} q_m^4. \tag{6}$$

Note that m_0 describes the component of the flux linear in the charge. For modelling purposes the memristor defined by Eq. (6) shall be seen as the series between a linear resistor of value m_0 and a memristor with memristance

$$M(q_m) = m_1 q_m + m_2 q_m^2 + m_3 q_m^3, \tag{7}$$

flux $\varphi_m(q_m) = \frac{m_1}{2} q_m^2 + \frac{m_2}{3} q_m^3 + \frac{m_3}{4} q_m^4$, and voltage $v_m = M(q_m)i_m$. The flux-charge relationship of the original memristor may then be recast as $\tilde{\varphi}_m(q_m) = \varphi_m(q_m) + m_0 q_m$, while the expression for its memristance reduces to $\tilde{M}(q_m) = M(q_m) + m_0$, and the voltage across it may be expressed in terms of the current through it as $\tilde{v}_m = v_m + m_0 i_m$. With reference to the memristor in Fig. 1, let us express its memristance as defined in Eq. (7), and shift its series resistor of resistance m_0 inside the two-port through one of its output port terminals.

We are now ready to model each element from the class of one-memristor[1] circuits shown in Fig. 1. The Volterra series paradigm will be adopted to derive the mathematical description of any circuit from the class. The first step of our systematic approach is to identify an association between electrical circuit's physical quantities and system theoretic variables.

[1]It is worthy to point out that the study presented here further extends the work reported in [14], where only ideal memristors [12] were allowed in the circuits modelled through the Volterra series paradigm.

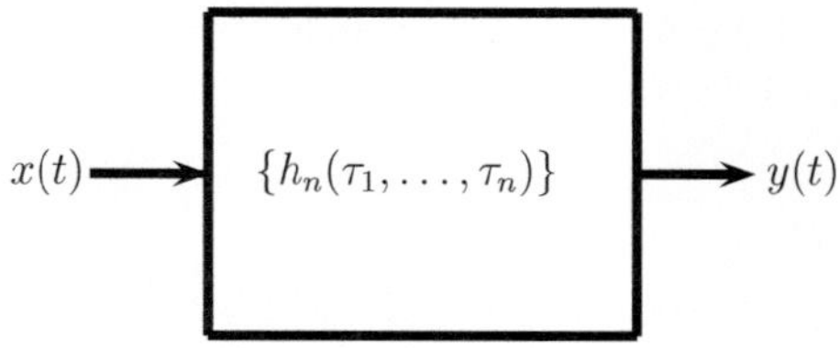

Fig. 2 System theoretic representation for the Volterra system $H[\cdot]$ identified in the circuit of Fig. 1

2.2 Application of the Theory to One-Memristor Circuits

In each circuit from the class of Fig. 1 a Volterra nonlinear dynamical system $H[\cdot]$, mapping a signal at the input port to a corresponding signal at the output port, may be identified. The system theoretic representation of this Volterra system, shown in Fig. 2, highlights input $x(t)$, and resulting output $y(t)$. The system is completely characterised by the set of Volterra kernels $\{h_n(\tau_1, \tau_2, , \ldots, \tau_n)\}$, where n ranges from 1 to ∞. The sufficient condition for a nth-order Volterra kernel $h_n(\tau_1, \ldots, \tau_n)$ to be Bounded Input Bounded Output (BIBO) stable is [10]

$$\int_{\tau_n=-\infty}^{\tau_n=\infty} \cdots \int_{\tau_1=-\infty}^{\tau_1=\infty} |h_n(\tau_1, \ldots, \tau_n)| d\tau_1 \ldots d\tau_n < \infty\,.$$

For a causal system, $h_n(\tau_1, \ldots, \tau_n) = 0$ for any n under $\tau_1 < 0$ and/or ... and/or $\tau_n < 0$ [10].

According to the nonlinear system theoretic paradigm under consideration [10], the response $y(t)$ of the system of Fig. 2 to a given input $x(t)$ may be approximated through a series expansion of Volterra operators arrested[2] to a certain order p, as follows:

$$y(t) \approx y^{(p)}(t) = H^{(p)}[x(t)] = \sum_{n=1}^{p} y_n(t), \tag{8}$$

where $H^{(p)}[x(t)]$ represents the pth-order approximation of the system response to the input $x(t)$, while $y_n(t) = H_n[x(t)]$ is the result of the application of the nth-order Volterra operator to the input, expressed in functional form as

$$y_n(t) = H_n[x(t)] = \int_{-\infty}^{+\infty} \cdots \int_{-\infty}^{+\infty} h_n(\tau_1, \ldots \tau_n) x(t-\tau_1) \ldots x(t-\tau_n) d\tau_1 \ldots d\tau_n. \tag{9}$$

[2]The larger is p, and the better is the degree of accuracy of the resulting Volterra series, which, in the limit $p \to \infty$, completely characterizes the system, i.e. $\lim_{p\to\infty} y^{(p)}(t) = y(t)$, in many cases.

Here $h_n(\tau_1, \ldots \tau_n)$ stands for the nth-order Volterra kernel [13], endowed with the following spectral representation:

$$H_n(j\omega_1, \ldots, j\omega_n) = \int_{-\infty}^{+\infty} \cdots \int_{-\infty}^{+\infty} h_n(\tau_1, \ldots \tau_n) e^{-(j\omega_1\tau_1 + \cdots + j\omega_n\tau_n)} d\tau_1 \ldots d\tau_n, \tag{10}$$

Once the set of the first p Volterra kernels of the system is obtained, Eqs. (8)–(9) may then be used to derive a pth-order approximation to the output response of the system under any excitation.

In order to accomplish the modelling task under consideration we first need to specify the association between physical quantities of the electrical circuit in Fig. 1 and system theoretic variables in Fig. 2. Let the input and output currents i_1 and i_2 of the two-port in the first figure represent the input and output x and y of the Volterra system of the latter figure, respectively. The analytical determination of the kernels of the Volterra system of Fig. 2 identified in the circuit of Fig. 1 requires the preliminary derivation of circuit-theoretic equations.

2.3 Circuit Dynamic Equations

We choose to describe the two-port in Fig. 1 through a current-voltage representation, as follows

$$i_1(t) = Y_{1,1}\{v_1(t)\} + Y_{1,2}\{v_2(t)\}, \quad \text{and} \tag{11}$$

$$i_2(t) = Y_{2,1}\{v_1(t)\} + Y_{2,2}\{v_2(t)\}. \tag{12}$$

The matrix whose elements are the integro-differential scalar operators $Y_{i,j}\{\cdot\}$ $(i, j \in 1, 2)$, reported in Eqs. (11)–(12), is defined as $\mathbf{Y}\{\cdot\}$, and expressed as

$$\mathbf{Y}\{\cdot\} = \begin{pmatrix} Y_{1,1}\{\cdot\} & Y_{1,2}\{\cdot\} \\ Y_{2,1}\{\cdot\} & Y_{2,2}\{\cdot\} \end{pmatrix}. \tag{13}$$

Let us define a scalar operator $T(\cdot)$, acting on the integro-differential scalar operator matrix $\mathbf{Y}\{\cdot\}$ as $T(\mathbf{Y})\{\cdot\} = Y_{1,1}\{Y_{2,2}\{\cdot\}\} - Y_{2,1}\{Y_{1,2}\{\cdot\}\}$. Under the assumption $Y_{2,1}\{Y_{1,1}\{\cdot\}\} = Y_{1,1}\{Y_{2,1}\{\cdot\}\}$, the scalar operators $Y_{2,1}\{\cdot\}$ and $Y_{1,1}\{\cdot\}$ are then applied to Eqs. (11) and (12), respectively. The subtraction of the resulting equations leads to

$$Y_{2,1}\{i_1(t)\} - Y_{1,1}\{i_2(t)\} = -T(\mathbf{Y})\{v_2(t)\}. \tag{14}$$

With reference to Fig. 1, using $i_m = -i_2$ and Eq. (7), the output port voltage $v_2(t)$ may be expressed as

$$v_2(t) = M(q_m(t))i_m(t) = -M(-q_2(t))i_2(t) = -M\left(-\int_{-\infty}^{t} i_2(t')dt'\right)i_2(t)$$
$$= -i_2(t)\sum_{l=1}^{3} m_l\left(-\int_{-\infty}^{t} i_2(t')dt'\right)^l. \tag{15}$$

Letting $x = i_1$, and $y = i_2$ [14] into Eq. (14), and using Eq. (15) gives

$$Y_{2,1}\{x(t)\} - Y_{1,1}\{y(t)\} = T(\mathbf{Y})\left\{y(t)\sum_{l=1}^{3} m_l\left(-\int_{-\infty}^{t} y(t')dt'\right)^l\right\}, \tag{16}$$

where $y(t)$, the output of the Volterra system of Fig. 2, is approximated through a pth-order series expansion, i.e. $y^{(p)}(t)$, as given in Eq. (8). The next section explains how Eq. (16), known as *dynamic equation* for the circuit class of Fig. 1, may be used to derive the *Volterra system kernels*.

2.4 Analytical Derivation of Volterra Kernels

The *harmonic probe technique* [15] is a useful theoretic methodology for the determination of the Volterra kernels of a nonlinear dynamical system in the spectral domain. For the determination of the Volterra kernel of order m (note that all the kernels of lower order need to be computed first), a m-tone complex exponential signal is applied to the input of the Volterra system of Fig. 2. Writing this input as

$$x(t) = \sum_{p=1}^{m} x_p(t), \tag{17}$$

with

$$x_p(t) \triangleq c e^{j\omega_p t} \tag{18}$$

($c \in \mathbb{R}_+$), and inserting it into Eq. (9), the corresponding Volterra system response arrested to the pth-order is found to be given by Eq. (8) with

$$y_n(t) = H_n[x(t)] = c^n \sum_{p_1=1}^{m} \cdots \sum_{p_n=1}^{m} H_n(j\omega_{p_1}, \ldots, j\omega_{p_n}) e^{(j\omega_{p_1}+\ldots+j\omega_{p_n})t}. \tag{19}$$

At this point the series expansions for input, as expressed by Eqs. (17)–(18) and output, as given in Eq. (19), are inserted into the dynamic equation (16) for the circuit class under modelling. Not all terms of these two series expansions need to be taken into account since computations within Eq. (16) are only allowed to deliver results containing factor c^m. Furthermore, within the resulting dynamic equation, only expressions proportional to $e^{(j\omega_1+j\omega_2+\ldots+j\omega_m)t}$ are retained, since only the mth-order kernel transform multiplied to $e^{(j\omega_1+j\omega_2+\ldots+j\omega_m)t}$ has distinct arguments $j\omega_1, j\omega_2, \ldots, j\omega_m$. The final equation is then solved for $H_m(j\omega_1, j\omega_2, \ldots, j\omega_m)$. The solution is found to depend upon lower order kernels, determined earlier in this systematic procedure. In order to derive the simplest expression for the mth-order kernel, the following analysis assumes that $H_n(j\omega_1, j\omega_2, \ldots, j\omega_n)$, i.e. the system kernel of order n, for all $n \in \{1, 2, \ldots, m\}$, is symmetric, i.e. the order of appearance of the variables $j\omega_1, j\omega_2, \ldots, j\omega_n$ within its argument, is irrelevant. This assumption is by no means restrictive. In fact it may be proved [10] that a nth-order Volterra operator $H_n[\cdot]$ may be equivalently characterized by any of its n asymmetric kernels $H_{n,asym}(\cdot, \cdot, \ldots, \cdot)$ or by its symmetrized kernel $H_n(j\omega_1, j\omega_2, \ldots, j\omega_n)$, defined as

$$H_n(j\omega_1, j\omega_2, \ldots, j\omega_n) = \frac{1}{n!} \sum_{\substack{all\ possible\ rearrangements\\ of\ the\ n\ arguments}} H_{n,asym}(\cdot, \cdot, \ldots, \cdot).$$

The next section determines a closed form expression for the 1st-order kernel of the Volterra system under study.

2.4.1 Kernel of Order 1

With $m = 1$, Eqs. (17)–(18) give $x(t) = ce^{j\omega_1 t}$. Further, Eq. (19) yields

$$y_n(t) = c^n H_n(j\omega_1, \ldots, j\omega_1)e^{jn\omega_1 t} \tag{20}$$

The dynamic equation (16) of the circuit class under study is allowed to contain terms with factor c only. Therefore, with reference to such equation, in the left hand side we replace $x(t)$ with $ce^{j\omega_1 t}$ and $y(t)$ with $c^1 H_1(j\omega_1)e^{j\omega_1 t}$, while the right hand side is set to 0. The dynamic equation reduces to

$$Y_{2,1}\{e^{j\omega_1 t}\} - Y_{1,1}\{e^{j\omega_1 t}\}H_1(j\omega_1) = 0. \tag{21}$$

From properties of integro-differential operators applied to complex exponential signals, we have

$$Y_{i,j}\{e^{j\omega_1 t}\} = Y_{i,j}(j\omega_1)e^{j\omega_1 t}, \tag{22}$$

for $i, j \in \{1, 2\}$, where $Y_{i,j}(j\omega_1)$ denotes the (i, j) element of the admittance matrix $\mathbf{Y}(j\omega)$, arranged as

$$\mathbf{Y}(j\omega) = \begin{pmatrix} Y_{1,1}(j\omega) & Y_{1,2}(j\omega) \\ Y_{2,1}(j\omega) & Y_{2,2}(j\omega) \end{pmatrix}. \tag{23}$$

The determinant of this matrix is defined as $\det(\mathbf{Y}(j\omega)) = Y_{1,1}(j\omega)Y_{2,2}(j\omega) - Y_{1,2}(j\omega)Y_{2,1}(j\omega)$. Using Eq. (22) into (21) finally gives the following expression for the kernel of order 1:

$$H_1(j\omega_1) = \frac{Y_{2,1}(j\omega_1)}{Y_{1,1}(j\omega_1)}, \tag{24}$$

The first-order kernel depends only upon properties of the two-port of Fig. 1 through the elements in the first column of the admittance matrix of Eq. (23).

2.4.2 Kernel of Order 2

Here we let $m = 2$. As a result Eqs. (17)–(18) deliver $x(t) = ce^{j\omega_1 t} + ce^{j\omega_2 t}$, while from Eq. (19) it follows that

$$y_n(t) = c^n \sum_{p_1=1}^{2} \cdots \sum_{p_n=1}^{2} H_n(j\omega_{p_1}, \ldots, j\omega_{p_n}) e^{(j\omega_{p_1}+\ldots+j\omega_{p_n})t} \tag{25}$$

From this equation the expressions for the Volterra operators of order 1 and 2 are found to be

$$y_1(t) = H_1[x(t)] \; = ce^{j\omega_1 t} H_1(j\omega_1) + ce^{j\omega_2 t} H_1(j\omega_2), \tag{26}$$

$$\begin{aligned} y_2(t) = H_2[x(t)] \; &= c^2 e^{2j\omega_1 t} H_2(j\omega_1, j\omega_1) + c^2 e^{2j\omega_2 t} H_2(j\omega_2, j\omega_2) \\ &+ c^2 \left(H_2(j\omega_1, j\omega_2) + H_2(j\omega_2, j\omega_1)\right) e^{j(\omega_1+\omega_2)t}. \end{aligned} \tag{27}$$

There is no need to compute higher-order Volterra operators since they would be proportional to powers of c larger than 2. Here the dynamic equation (16) may only contain terms with factor c^2. As a result, on its left hand side $Y_{21}\{x(t)\} = cY_{21}e^{(j\omega_1 t)} + e^{(j\omega_2 t)}$ may not provide any contribution, and substitute $y(t)$ with $y_2(t)$, as expressed in Eq. (27). Regarding the right hand side of the dynamic equation, l is taken equal to 1, while $y(t)$ is replaced with $y_1(t)$, as reported in Eq. (26), in both its appearances. The dynamic equation (16) may then be cast as

$$-Y_{11}\{y_2(t)\} = -T(\mathbf{Y}) \left\{ y_1(t) m_1 \left(\int_{-\infty}^{t} y_1(t')dt' \right) \right\}. \tag{28}$$

Inserting Eqs. (26)–(27) into this, computing the integral, and retaining only expressions proportional to $e^{j(\omega_1+\omega_2)t}$, the following equation is obtained:

$$\begin{aligned} -Y_{11}\left\{e^{j(\omega_1+\omega_2)t}\right\}\left(H_2(j\omega_1, j\omega_2) + H_2(j\omega_2, j\omega_1)\right) &= -m_1 T(\mathbf{Y}) \left\{e^{j(\omega_1+\omega_2)t}\right\} \\ &\cdot \left(\frac{j\omega_1 + j\omega_2}{j\omega_1 j\omega_2}\right) \prod_{i=1}^{2} H_1(j\omega_i) \end{aligned} \tag{29}$$

Making use of the identity established in Eq. (22), and using the symmetric property of second-order kernels, $H_2(j\omega_1, j\omega_2)$ may be finally expressed as follows:

$$H_2(j\omega_1, j\omega_2) = \frac{m_1}{j\omega_1} \frac{\det(\mathbf{Y}(j(\omega_1+\omega_2)))}{Y_{11}(j(\omega_1+\omega_2))} H_1(j\omega_1) H_1(j\omega_2). \tag{30}$$

With reference to the circuit class under modeling (see Fig. 1), it is instructive to observe that the frequency-domain expression for the second-order kernel depends not only upon the properties of the two-port through the admittance matrix of Eq. (23), but also upon the memristor load via coefficient m_1 of the memristance function defined in Eq. (7).

2.4.3 Kernel of Order 3

We now set $m = 3$. Equations (17)–(18) reduce to $x(t) = ce^{j\omega_1 t} + ce^{j\omega_2 t} + ce^{j\omega_3 t}$, while Eq. (19) delivers

$$y_n(t) = c^n \sum_{p_1=1}^{3} \cdots \sum_{p_n=1}^{3} H_n(j\omega_{p_1}, \ldots, j\omega_{p_n}) e^{(j\omega_{p_1}+\ldots+j\omega_{p_n})t} \tag{31}$$

From this equation, taking into account the symmetric property of the third-order kernel, the expressions for the Volterra operators of order 1, 2, and 3 are found to be

$$y_1(t) = c^1 \left(H_1(j\omega_1)e^{j\omega_1 t} + H_1(j\omega_2)e^{j\omega_2 t} + H_1(j\omega_3)e^{j\omega_3 t} \right) \tag{32}$$

$$\begin{aligned} y_2(t) = c^2 \big(& H_2(j\omega_1, j\omega_1)e^{2j\omega_1 t} + H_2(j\omega_2, j\omega_2)e^{2j\omega_2 t} + H_2(j\omega_3, j\omega_3)e^{2j\omega_3 t} \\ & + 2H_2(j\omega_1, j\omega_2)\, e^{j(\omega_1+\omega_2)t} + 2H_2(j\omega_1, j\omega_3)e^{j(\omega_1+\omega_3)t} \\ & + 2H_2(j\omega_2, j\omega_3)e^{j(\omega_2+\omega_3)t} \big) \end{aligned} \tag{33}$$

$$\begin{aligned} y_3(t) = c^3 \Big(& H_3(j\omega_1, j\omega_1, j\omega_1)e^{3j\omega_1 t} + H_3(j\omega_2, j\omega_2, j\omega_2)e^{3j\omega_2 t} \\ & + H_3(j\omega_3, j\omega_3, j\omega_3)e^{3j\omega_3 t} + 6H_3(j\omega_1, j\omega_2, j\omega_3)e^{j(\omega_1+\omega_2+\omega_3)t} \\ & + 3H_3(j\omega_1, j\omega_1, j\omega_2)e^{j(2\omega_1+\omega_2)t} + 3H_3(j\omega_1, j\omega_2, j\omega_2)e^{j(\omega_1+2\omega_2)t} \\ & + 3H_3(j\omega_1, j\omega_1, j\omega_3)e^{j(2\omega_1+\omega_3)t} + 3H_3(j\omega_1, j\omega_3, j\omega_3)e^{j(\omega_1+2\omega_3)t} \\ & + 3H_3(j\omega_2, j\omega_2, j\omega_3)e^{j(2\omega_2+\omega_3)t} + 3H_3(j\omega_2, j\omega_3, j\omega_3)e^{j(\omega_2+2\omega_3)t} \Big) \end{aligned} \tag{34}$$

Higher-order Volterra operators are proportional to powers of c larger than 3. As a result, their computation is meaningless, since the dynamic equation (16) may only contain terms with factor c^3. On the left hand side of Eq. (16), $Y_{21}\{x(t)\} = cY_{21}e^{(j\omega_1 t)} + e^{(j\omega_2 t)} + e^{(j\omega_3 t)}$ may not provide any contribution, while $y(t)$ is replaced with $y_3(t)$, described by Eq. (34). Regarding the right hand side of Eq. (16), it delivers three additive terms proportional to c^3, two for $l = 1$, and one for $l = 2$. In the first addend associated with $l = 1$ the first and second copies of $y(t)$ are respectively substituted by $y_1(t)$, defined in Eq. (32), and $y_2(t)$, expressed by Eq. (33). The second term obtained for $l = 1$ has $y_2(t)$ and $y_1(t)$ in place for first and second copies of $y(t)$. Finally, the $l = 2$ expression has $y_1(t)$ for both occurrences of $y(t)$. The dynamic equation (16) may then be cast as

$$T(\mathbf{Y}) \left\{ \left(m_1 \left(\int_{-\infty}^{t} y_2(t')dt' \right) - m_2 \left(\int_{-\infty}^{t} y_1(t')dt' \right)^2 \right) y_1(t) + m_1 y_2(t) \right.$$
$$\left. \cdot \left(\int_{-\infty}^{t} y_1(t')dt' \right) \right\} = Y_{1,1}\{y_3(t)\}, \tag{35}$$

Inserting Eqs. (32)–(34) into this, computing the integrals, retaining only expressions proportional to $e^{j(\omega_1+\omega_2+\omega_3)t}$, and taking Eq. (22) into account, the third-order kernel is found to be given by

$$H_3(j(\omega_1, \omega_2, \omega_3)) = \frac{\det(\mathbf{Y}(j(\omega_1+\omega_2+\omega_3)))}{3\hat{Y}_{11}(j(\omega_1+\omega_2+\omega_3))} \left\{ m_1 H_1(j\omega_3) H_2(j\omega_1, j\omega_2) \right.$$
$$\cdot \left(\frac{1}{j\omega_3} + \frac{1}{j(\omega_1+\omega_2)} \right) + m_1 H_1(j\omega_2) H_2(j\omega_1, j\omega_3)$$
$$\cdot \left(\frac{1}{j\omega_2} + \frac{1}{j(\omega_1+\omega_3)} \right) + m_1 H_1(j\omega_1) H_2(j\omega_2, j\omega_3)$$
$$\cdot \left(\frac{1}{j\omega_1} + \frac{1}{j(\omega_2+\omega_3)} \right) - m_2 H_1(j\omega_1) H_1(j\omega_2) H_1(j\omega_3)$$
$$\left. \cdot \left(\frac{1}{j\omega_1 j\omega_2} + \frac{1}{j\omega_1 j\omega_3} + \frac{1}{j\omega_2 j\omega_3} \right) \right\}. \tag{36}$$

The spectral domain representation of the third-order kernel depends more strongly than the one pertaining to the second-order kernel (compare Eqs. (30) and (36)) upon the memristance parameters of the memristor of Fig. 1.

A similar procedure, as described in Sects. 2.4.1, 2.4.2, and 2.4.3, allows the derivation of higher-order kernels. With reference to Eq. (48), the order $n = p$ at which arresting the determination of the system kernel transforms $H_n(j\omega_1, \ldots, j\omega_n)$ depends upon the required level of accuracy of the Volterra series-based model.

2.5 *Response to a Sine-Wave Input*

Under a certain input $x(t)$, knowledge of all the Volterra kernels up to the pth order allows the computation of the pth-order approximation $y^{(p)}(t)$ of the system response $y(t)$ through the use of Eqs. (8)–(9). In these equations the kernels appear in the time domain, while in Sects. 2.4.1, 2.4.2, and 2.4.3 they were derived in the spectral domain. The time-domain kernel $h_m(\tau_1, \tau_2, \ldots, \tau_m)$ may be obtained from its spectral domain representation $H_m(j\omega_1, j\omega_2, \ldots, j\omega_m)$ through

$$h_m(\tau_1, \ldots \tau_m) = \int_{-\infty}^{+\infty} \cdots \int_{-\infty}^{+\infty} \frac{H_m(j\omega_1, \ldots, j\omega_m)}{(2\pi)^m} e^{(j\omega_1\tau_1 + \cdots + j\omega_m\tau_m)} d\omega_1 \ldots d\omega_m \,. \quad (37)$$

However, there is an alternative approach which allows a direct use of the spectral domain kernels for the computation of the approximated system response $y^{(p)}(t)$ in Eq. (8). Without loss of generality, let us assume that the input to the system is a sine wave expressed by $x(t) = x_0 \sin \omega_0 t = \frac{1}{2j} \sum_{i=-1, i \neq 0}^{+1} x_i \text{sign}(i) e^{j\omega_i t}$ where $x_i = x_0$ for $i \in \{-1, 1\}$, $\omega_1 = -\omega_{-1} = \omega_0$, and sign$(\cdot)$ stands for the signum function. Inserting this into Eq. (9), and manipulating the resulting expression, the nth-order Volterra series of $y^{(p)}(t)$ is found to be directly related to the spectral domain representation of the nth-order Volterra kernel [16], assuming the following form:

$$y_n(t) = \sum_{i_1=-1,\ i_1 \neq 0}^{+1} \cdots \sum_{i_n=-1,\ i_n \neq 0}^{+1} \frac{H_n(j\omega_{i_1}, \ldots j\omega_{i_n})}{(2j)^n} e^{j(\omega_{i_1} + \ldots + \omega_{i_n})t} \prod_{k=1}^{n} x_{i_k} \text{sign}(i_k). \quad (38)$$

Let us derive the expressions this equation assumes for $n \in \{1, 2, 3\}$ using properties of trigonometric functions and complex numbers. Let $\Re\{\cdot\}$ and $\Im\{\cdot\}$ denote the operations extracting real and imaginary part from a complex argument, respectively. Further * represents the complex conjugate operator. Using equality

$$H_1(-j\omega_0)e^{-j\omega_0 t} = \left(H_1(j\omega_0)e^{j\omega_0 t}\right)^*,$$

for $n = 1$ Eq. (38) delivers

$$y_1(t) = x_0 \Im\{H_1(j\omega_0)e^{j\omega_0 t}\}. \quad (39)$$

Employing relations

$$\begin{aligned} H_2(-j\omega_0, -j\omega_0)e^{-2j\omega_0 t} &= \left(H_2(j\omega_0, j\omega_0)e^{2j\omega_0 t}\right)^*, \\ H_2(j\omega_0, -j\omega_0) &= \left(H_2(-j\omega_0, +j\omega_0)\right)^*, \\ \Re\{H_2(j\omega_0, -j\omega_0)\} &= H_2(j\omega_0, -j\omega_0), \end{aligned}$$

the latter descending from the symmetric property of the second-order kernels, for $n = 2$ Eq. (38) yields

$$y_2(t) = -\frac{x_0^2}{2} \Big(-\Re\{H_2(j\omega_0, -j\omega_0)\} + \Re\{H_2(j\omega_0, j\omega_0)e^{2j\omega_0 t}\} \Big). \quad (40)$$

Finally, taking into account the symmetric property of the third-order kernels, and making use of identity

$$H_3(-j\omega_0, -j\omega_0, -j\omega_0)e^{-3j\omega_0 t} = \left(H_3(j\omega_0, j\omega_0, j\omega_0)e^{3j\omega_0 t}\right)^*,$$

the expression obtained from Eq. (38) for $n = 3$ is

$$y_3(t) = -\frac{x_0^3}{4}\left(\Im\{H_3(j\omega_0, j\omega_0, j\omega_0)e^{3j\omega_0 t}\} + \Im\left\{\left(H_3(-j\omega_0, j\omega_0, j\omega_0) + H_3(j\omega_0, -j\omega_0, j\omega_0) + H_3(j\omega_0, j\omega_0, -j\omega_0)\right) e^{j\omega_0 t}\right\}\right). \quad (41)$$

Higher-order Volterra operators may be derived in a similar manner from Eq. (38). Finally, the sum of all the Volterra operators up to the pth order, according to Eq. (8), provides the desired expression for the pth-order approximation $y^{(p)}(t)$ of the system response $y(t)$ to the sine wave input $x(t) = x_0 \sin \omega_0 t$.

The next section presents a couple of case studies of an illustrative circuit from the class of Fig. 1 to elucidate the applicability of the Volterra series approach to the modelling of the one-memristor circuits under exam.

2.6 Theory Validation

Let us consider the circuit of Fig. 3, where the dashed rectangular box contains a particular linear and dynamic two-port consisting of resistors, capacitors, and inductors. As expected from Sect. 2.1, the series resistor of the memristor is shifted into the two-port through its upper output terminal. Application of standard circuit theoretic principles to this electrical system delivers the following state equations:

$$\begin{aligned}
\frac{dy_1}{dt} &= \frac{1}{C_1}\left(i_1 - y_2 - \frac{v_1}{R_1}\right),\\
\frac{dy_2}{dt} &= \frac{1}{L_1} y_1,\\
\frac{dy_3}{dt} &= \frac{1}{L_2}\left(-y_4 + (M(q_m) + m_0)i_m\right),\\
\frac{dy_4}{dt} &= -\frac{1}{C_2} i_m,
\end{aligned} \quad (42)$$

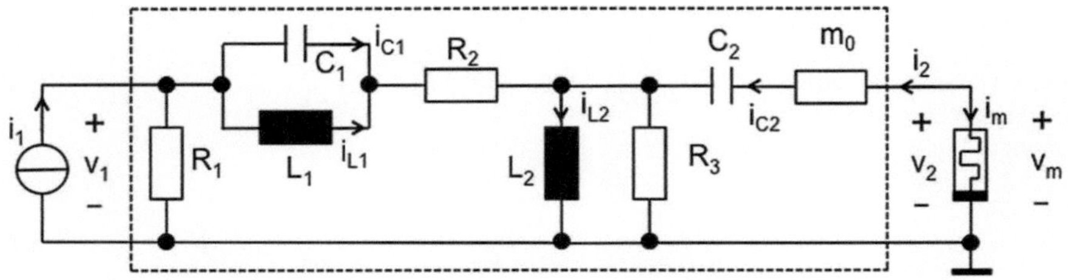

Fig. 3 Circuit example. The two-port is enclosed within the dashed rectangular box

where

$$\begin{aligned} q_m &= -C_2 y_4, \\ i_m &= \frac{-\alpha y_3 + (R_3^{-1} + R_1^{-1}\beta) y_4 - R_1^{-1} y_1 + i_1}{(M(q_m) + m_0)(R_3^{-1} + R_1^{-1}\beta) + \alpha}, \\ v_m &= y_1 + R_2 y_3 - \beta y_4 + (M(q_m) + m_0)\beta i_m, \end{aligned}$$

$\alpha = 1 + \frac{R_2}{R_1}$, $\beta = 1 + \frac{R_2}{R_3}$, and the state variables are defined as $y_1 \triangleq v_{C_1}$, $y_2 \triangleq i_{L_1}$, $y_3 \triangleq i_{L_2}$, and $y_4 \triangleq v_{C_2}$, where v_{C_i} and i_{L_i} represent voltage across capacitor C_i and current through inductor L_i, respectively ($i \in \{1, 2\}$).

The admittance matrix of the 4th-order non autonomous system, necessary for the computation of the Volterra theory-based time series of all currents and voltages within the circuit of Fig. 3, is given by Eq. (23). The elements of this matrix, expressed in the Laplace domain through $s = j\omega$, are found to be expressed by

$$\begin{aligned} Y_{11}(s) &= Y_1(s) + R_1^{-1}, \\ Y_{12}(s) &= -\left(R_2 + \left(sC_1 + s^{-1}L_1^{-1}\right)^{-1}\right)^{-1} Y_2(s)^{-1} Y_{22}(s), \\ Y_{21}(s) &= -\left(m_0 + s^{-1}C_2^{-1}\right)^{-1}\left(s^{-1}L_2^{-1} + R_3^{-1} + \left(m_0 + s^{-1}C_2^{-1}\right)^{-1}\right)^{-1} \\ &\quad \cdot \frac{R_1}{R_1 + Y_1^{-1}(s)} Y_{11}(s), \\ Y_{22}(s) &= \left(Y_2^{-1}(s) + m_0 + s^{-1}C_2^{-1}\right)^{-1}, \end{aligned}$$

where $Y_1(s)$ and $Y_2(s)$ have the following forms:

$$Y_1(s) = \left(\left(\frac{sC_2}{1 + sC_2 m_0} + R_3^{-1} + s^{-1}L_2^{-1}\right)^{-1} + R_2 + \frac{1}{sC_1 + s^{-1}L_1^{-1}}\right)^{-1}, \text{ and}$$

$$Y_2(s) = \left(R_2 + \left(sC_1 + s^{-1}L_1^{-1}\right)^{-1}\right)^{-1} + s^{-1}L_2^{-1} + R_3^{-1}.$$

The remaining part of this manuscript gives evidence for the qualitative and quantitative agreement between the numerical solutions to the circuit Eq. (42) and the

Table 1 Parameter setting for the first case study of the circuit of Fig. 3

$R_1/k\Omega$	$C_1/\mu F$	L_1/mH	R_2/Ω
10	2	5	100
L_2/mH	R_3/Ω	$C_2/\mu F$	m_0/Ω
0.1	10	100	20
$m_1/(\Omega\, C^{-1})$	$m_2/(\Omega\, C^{-2})$	$m_3/(\Omega\, C^{-3})$	$\mathbf{y}(0)$
10^7	10^2	$5 \cdot 10^4$	$\mathbf{0}$

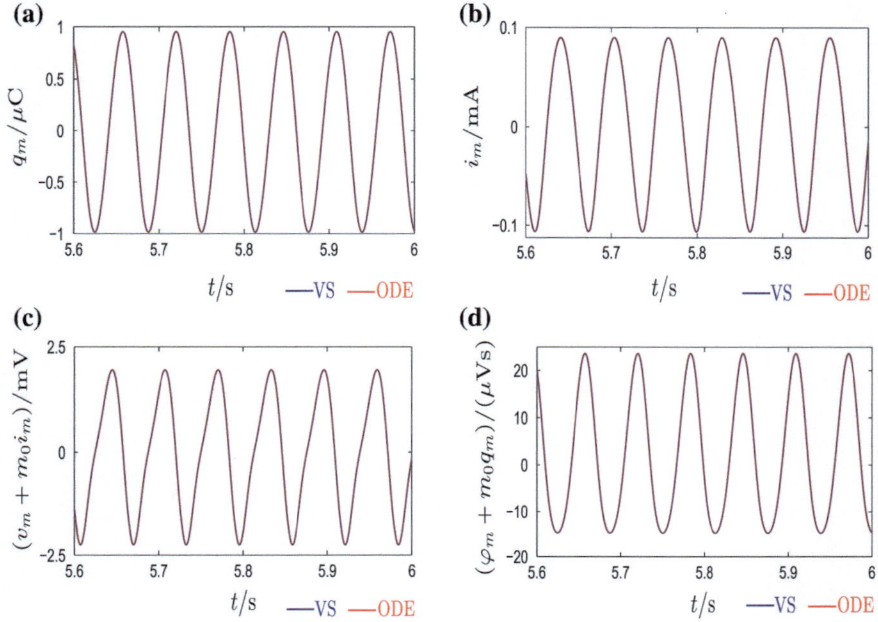

Fig. 4 First set of simulation results for the first case study. Time waveforms of q_m (**a**), i_m (**b**), $\tilde{v}_m = v_m + m_0 i_m$ (**c**), and $\tilde{\varphi}_m(q_m) = \varphi_m(q_m) + m_0 q_m$ (**d**)

predictions of the Volterra series-based model, developed in Sect. 2.4, for a couple of case studies of the 4th-order non-autonomous electronic system of Fig. 3. In both the case studies this circuit is excited by a sine input $i_1(t) = x(t) = x_0 \sin \omega_0 t$ with amplitude $x_0 = 1$A and angular frequency $\omega_0 = 100\frac{\text{rad}}{\text{s}}$. In all the simulation results to follow red and blue waveforms shall refer to the numerical solutions of the Ordinary Differential Equations (ODE) (42) and to the corresponding Volterra series (VS)-based predictions.

For the first case study Table 1 reports the parameter values of the circuit of Fig. 3 as well as the initial condition $\mathbf{y}(0) = [y_1(0), y_2(0), y_3(0), y_4(0)]'$ for the numerical integration of equations (42).

Here the largest coefficient in the memristance expression of Eq. (5) is m_1.

As anticipated in Sect. 2.1, despite the separation between linear and nonlinear parts of the memristor flux-charge relation, previously adopted for modelling purposes, the memristance of the memristor is actually expressed as $\tilde{M}(q_m) = m_0 + M(q_m)$, as given in Eq. (5), where $M(q_m)$ is reported in Eq. (7).

Figure 4 depicts the time series of memristor charge q_m (plot (a)), current i_m (plot (b)), voltage $\tilde{v}_m = v_m + m_0 i_m$ (plot (c), $v_m = M(q_m) i_m$) and flux $\tilde{\varphi}_m(q_m) = \varphi_m(q_m) + m_0 q_m$ (plot (d), $\varphi_m(q_m) = \frac{m_1}{2} q_m^2 + \frac{m_2}{3} q_m^3 + \frac{m_3}{4} q_m^4$).

Figure 5a shows the power spectrum of the memristor current, while Fig. 5b depicts the steady-state pinched hysteresis loop emerging in the $\tilde{v}_m = v_m + m_0 i_m$-$i_m$ plane. The top and bottom curves in plot (c) of Fig. 5 correspond to the charge dependence

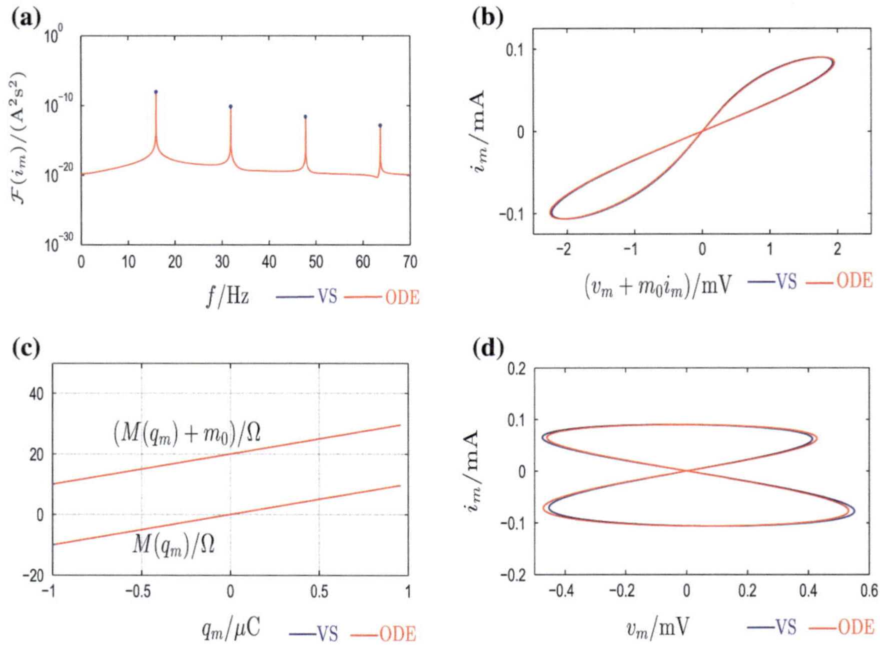

Fig. 5 Second set of simulation results for the first case study. **a** Power spectrum of the memristor current. **b** Steady-state pinched hysteresis loop in the $\tilde{v}_m = v_m + m_0 i_m$-$i_m$ plane ($v_m = M(q_m)i_m$). **c** $\tilde{M}(q_m) = M(q_m) + m_0$ (top curve) and $M(q_m)$ (bottom curve) versus charge q_m. **d** Steady-state pinched hysteresis loop in the v_m-i_m plane ($v_m = M(q_m)i_m$)

Table 2 Parameter values for the second case study of the fourth-order non autonomous system

$R_1/k\Omega$	C_1/μF	L_1/mH	R_2/Ω
10	2	5	100
L_2/mH	R_3/Ω	C_2/μF	m_0/Ω
0.1	10	100	20
$m_1/(\Omega\, C^{-1})$	$m_2/(\Omega\, C^{-2})$	$m_3/(\Omega\, C^{-3})$	$\mathbf{y}(0)$
10^7	10^{13}	$5 \cdot 10^4$	$\mathbf{0}$

of $\tilde{M}(q_m) = M(q_m) + m_0$ and of $M(q_m)$. The latter goes negative in the course of the simulation. This interesting aspect implies that the steady-state pinched hysteresis loop observed in the v_m-i_m plane visits also second and fourth quadrants, a clear indication of *active behaviour*.

Table 2 specifies the parameter setting for the second case study. The only difference lies in the value of m_2, which is here the largest coefficient in the memductance expression of Eq. (5).

Plots (a), (b), (c), and (d) of Fig. 6 show the time evolution of q_m, i_m, $\tilde{v}_m = v_m + m_0 i_m$, and $\tilde{\varphi}_m = \varphi_m(q_m) + m_0 q_m$.

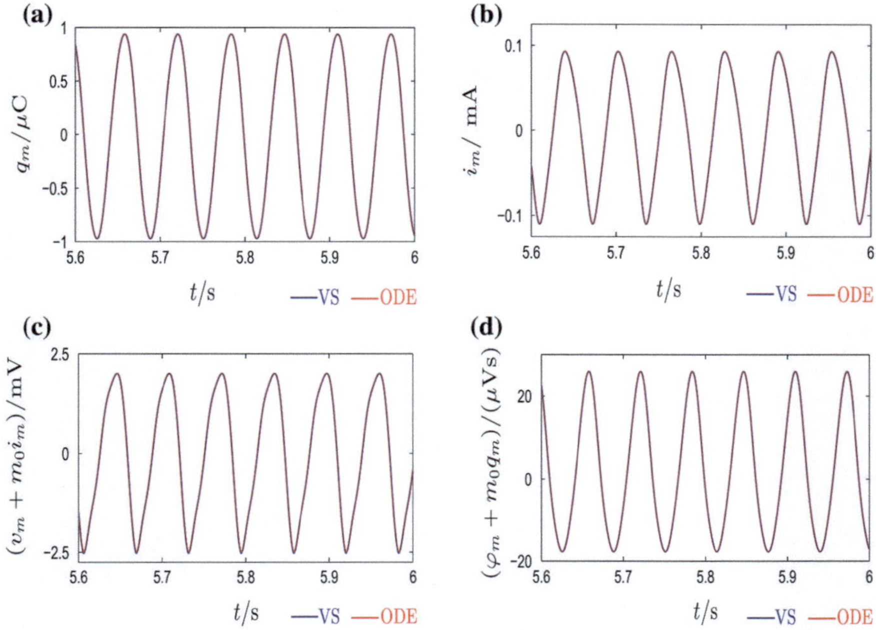

Fig. 6 First set of simulation results for the second case study. Time waveforms of q_m (**a**), i_m (**b**), $\tilde{v}_m = v_m + m_0 i_m$ (**c**), and $\tilde{\varphi}_m(q_m) = \varphi_m(q_m) + m_0 q_m$ (**d**)

Finally Fig. 7a–d respectively illustrate the power spectrum of i_m, the steady-state pinched hysteresis loop in the $\tilde{v}_m$-i_m plane, the plots of $\tilde{M}(q_m) = M(q_m) + m_0$ (top curve) and $M(q_m)$ (bottom curve) versus q_m, and the steady-state pinched hysteresis loop in the v_m-i_m plane. In the latter plot the steady-state loop goes through second and fourth quadrants as a result of the negative sign assumed periodically by the memductance function $M(q_m)$ in the course of the simulation, as clear from Fig. 7c. In conclusion, the description of the two case studies of the illustrative circuit of Fig. 3 has given evidence for the accuracy of the proposed nonlinear system theory-based modeling technique.

All in all, this section has applied the Volterra series paradigm to the mathematical characterization of the class of one-memristor circuits shown in Fig. 1. The proposed modelling approach may be generalized for the description of multi-memristor circuits. As a proof-of-concept, the next section presents the Volterra theory-based mathematical representation of a class of two-memristor circuits.

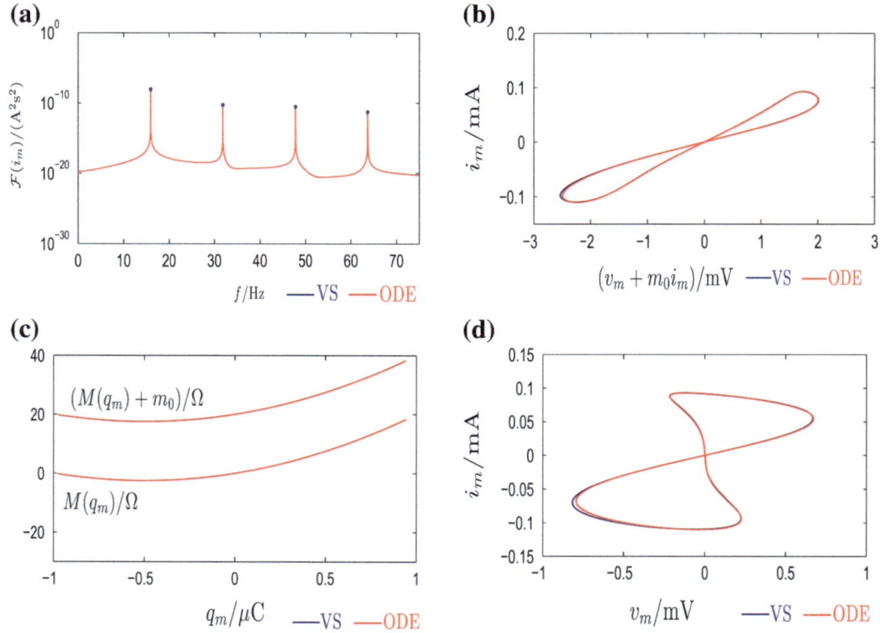

Fig. 7 Second set of simulation results for the second case study. **a** Power spectrum of the memristor current. **b** Steady-state pinched hysteresis loop in the $\tilde{v}_m = v_m + m_0 i_m$-$i_m$ plane ($v_m = M(q_m) i_m$). **c** $\tilde{M}(q_m) = M(q_m) + m_0$ (top curve) and $M(q_m)$ (bottom curve) versus charge q_m. **d** Steady-state pinched hysteresis loop in the v_m-i_m plane ($v_m = M(q_m) i_m$)

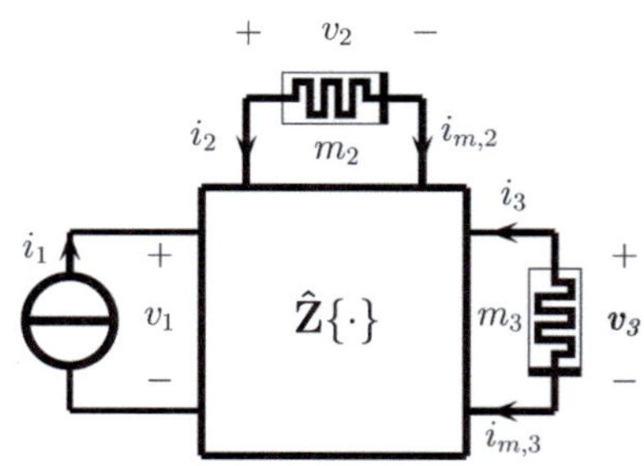

Fig. 8 Class of two-memristor circuits under modelling. $\mathbf{Z}\{\cdot\}$ denotes the matrix of integro-differential operators in the voltage-current description of the three-port

3 Extension to Two-Memristor Circuits

With reference to Fig. 8, the linear and dynamic three-ports under focus in this section have port 1 driven by a current source and each of ports 2 and 3 closed on a memristor.

As in Sect. 2, each memristor in Fig. 8 is assumed to be *ideal generic* [11]. Details on memristor modelling are provided in the next section.

3.1 Memristor Model

With reference to the rth memristor ($r \in \{2, 3\}$, see Fig. 8), the ordinary differential equation for the state w_r and the Ohm's state-dependent law are defined as

$$\frac{dw_r}{dt} = f(w_r)i_{m,r} = 3w_r^{\frac{2}{3}} i_{m,r}, \tag{43}$$

$$\tilde{v}_{m,r} = \tilde{M}(w_r)i_{m,r} = \left(\sum_{k'=0}^{3} m_{r,k'} w_r^{\frac{k'}{3}}\right) i_{m,r} \tag{44}$$

where $i_{m,r}$ and $\tilde{v}_{m,r}$ denote current through and voltage across the rth memristor, $\tilde{M}(w_r)$ stands for its memristance, while $m_{r,0}, m_{r,1}, m_{r,2}, m_{r,3} \in \mathbb{R}_{0,+}$. The rth ideal generic memristor defined by Eqs. (43)–(44) is equivalent to an ideal memristor with memristance $\tilde{M}(q_{m,r})$ and flux $\tilde{\varphi}_{m,r}$–charge $q_{m,r}$ constitutive relation respectively given by

$$\tilde{M}(q_{m,r}) = \sum_{k'=0}^{3} m_{r,k'} q_{m,r}^{k'}, \tag{45}$$

$$\tilde{\varphi}_{m,r}(q_{m,r}) = \sum_{k'=0}^{3} \frac{m_{r,k'}}{k'+1} q_{m,r}^{k'+1}. \tag{46}$$

This equivalence holds because the charge through the rth memristor is a one-to-one differentiable function of its state [11], namely $q_{m,r} = w_r^{\frac{1}{3}}$ here. For modelling purposes, the rth memristor defined by Eq. (46) may be seen as the series combination between a linear resistor of resistance $m_{r,0}$ and a memristor m_r with memristance

$$M(q_{m,r}) = \sum_{k'=1}^{3} m_{r,k'}\, q_{m,r}^{k'}, \tag{47}$$

flux $\varphi_m(q_m) = \sum_{k'=1}^{3} \frac{m_{r,k'}}{k'+1} q_{m,r}^{k'+1}$, and voltage $v_{m,r} = M(q_{m,r})i_{m,r}$. It follows that, for the original memristor, the flux-charge relation may be recast as $\tilde{\varphi}_{m,r}(q_{m,r}) = \varphi_m(q_m) + m_{0,r}q_{m,r}$, the memristance expression reduces to $\tilde{M}(q_{m,r}) = M(q_{m,r}) + m_{r,0}$, and the voltage may expressed in terms of the current as $\tilde{v}_{m,r} = v_{m,r} + m_{0,r}i_m$. With reference to the memristor m_r ($r \in \{2, 3\}$) in Fig. 8, let us express its memristance as defined in Eq. (47), and shift its series resistor of resistance $m_{0,r}$ inside the three-port through one of the two output terminals at port r.

In the next sections the Volterra series paradigm shall be applied to the modelling of each element from the class of two-memristor circuits of Fig. 8. As first step in the proposed technique an association between electrical circuit's physical quantities and system theoretic variables has to be identified.

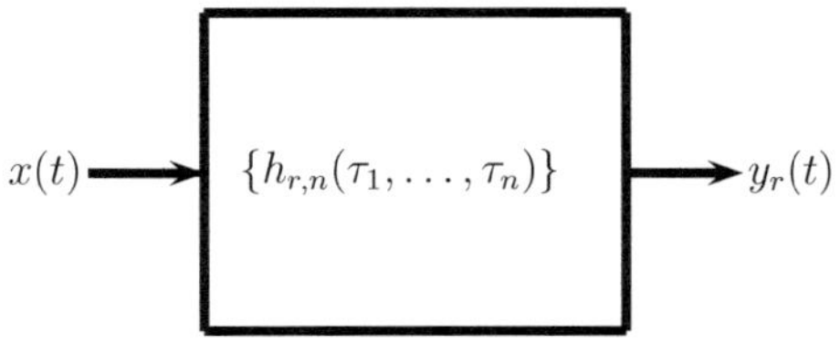

Fig. 9 System theoretic representation for the $(r-1)$th Volterra system $H_r[\cdot]$ identified in the circuit of Fig. 8 ($r \in \{2, 3\}$)

3.2 Application of the Theory to Two-Memristor Circuits

In each element from the class of two-memristor circuits in Fig. 8 two Volterra systems, mapping a signal at input port 1 to a signal at output port 2 or 3 respectively, may be identified. Let the output of Volterra system $H_r[\cdot]$ refer to a signal at port r ($r \in \{2, 3\}$). The system theoretic representation of the $(r-1)$th Volterra system is shown in Fig. 9.

The Volterra series-based theoretic representation of nonlinear dynamical system $H_r[\cdot]$ with input $x(t)$ and output $y_r(t)$ is the set of kernels $\{h_{r,n}(\tau_1, \ldots, \tau_n)\}$ for $n \in \{1, \ldots, \infty\}$. The sufficient condition for a nth-order Volterra kernel $h_{r,n}(\tau_1, \ldots, \tau_n)$ to be BIBO stable is [10]

$$\int_{\tau_n=-\infty}^{\tau_n=\infty} \cdots \int_{\tau_1=-\infty}^{\tau_1=\infty} |h_{r,n}(\tau_1, \ldots, \tau_n)| d\tau_1 \ldots d\tau_n < \infty .$$

For a causal system, $h_{r,n}(\tau_1, \ldots, \tau_n) = 0$ for any n under $\tau_1 < 0$ and/or ... and/or $\tau_n < 0$ [10]. In practical cases n is only swept up to a certain finite number p. The output of system $H_r[\cdot]$ ($r \in \{2, 3\}$) is approximated as a series expansion arrested to a certain order p, namely

$$y_r(t) \approx y_r^{(p)}(t) = H_r^{(p)}[x(t)] = \sum_{n=1}^{p} y_{r,n}(t), \tag{48}$$

where $H_r^{(p)}[x(t)]$ represents the pth-order approximation of the response of the $(r-1)$th system to the input $x(t)$, while $y_{r,n}(t) = H_{r,n}[x(t)]$ is the result of the application of the nth-order Volterra operator of system $H_r[\cdot]$ to the input, expressed in functional form as a n-dimensional convolution integral involving system input and nth-order Volterra kernel, namely

$$y_{r,n}(t) = H_{r,n}[x(t)] = \int_{-\infty}^{+\infty} \cdots \int_{-\infty}^{+\infty} h_{r,n}(\tau_1, \ldots \tau_n) x(t-\tau_1) \ldots x(t-\tau_n) d\tau_1 \ldots d\tau_n . \tag{49}$$

The nth-order Volterra kernel $h_{r,n}(\tau_1, \ldots \tau_n)$ appearing in Eq. (49) has the following spectral representation [10]:

$$H_{r,n}(j\omega_1,\ldots,j\omega_n) = \int_{-\infty}^{+\infty}\cdots\int_{-\infty}^{+\infty} h_{r,n}(\tau_1,\ldots\tau_n)e^{-(j\omega_1\tau_1+\cdots+j\omega_n\tau_n)}d\tau_1\ldots d\tau_n\,. \tag{50}$$

For the $(r-1)$th Volterra system $H_r[\cdot]$, an association between system theoretic variables (input $x(t)$, and output $y_r(t)$, see Fig. 9), and electrical circuit's physical quantities ($i_1(t)$, and $i_r(t) = -i_{m,r}(t)$, see Fig. 8) is then established in the given order ($r \in \{2, 3\}$).

The analytical determination of the kernels of each of the Volterra systems (Fig. 2 refers to system $H_r[\cdot]$, and $r \in \{2, 3\}$) identified in the circuit of Fig. 8 requires the preliminary derivation of circuit-theoretic equations.

3.3 Circuit Dynamic Equations

Without loss of generality, let us characterize the three-port in Fig. 8 through the voltage-current representation, namely

$$v_1(t) = Z_{1,1}\{i_1(t)\} + Z_{1,2}\{i_2(t)\} + Z_{1,3}\{i_3(t)\}, \tag{51}$$
$$v_2(t) = Z_{2,1}\{i_1(t)\} + Z_{2,2}\{i_2(t)\} + Z_{2,3}\{i_3(t)\}, \tag{52}$$
$$v_3(t) = Z_{3,1}\{i_1(t)\} + Z_{3,2}\{i_2(t)\} + Z_{3,3}\{i_3(t)\}. \tag{53}$$

where $Z_{k,l}\{\cdot\}$ denotes the element at row k and column l in the matrix of integro-differential operators $\mathbf{Z}\{\cdot\}$ ($k, l \in \{1, 2, 3\}$). Let us express the voltage v_r at port r ($r \in \{2, 3\}$) as $v_r = M(q_{m,r})i_{m,r}$, where $M(q_{m,r})$ is described by (47). Recalling the definitions for input and output of Volterra system $H_r[\cdot]$, namely $x(t) = i_1(t)$ and $y_r(t) = i_r(t)$, Eqs. (52) and (53) may be recast as

$$\begin{aligned} &-\left(\sum_{k'=1}^{3} m_{2,k'}\left(-\int_{-\infty}^{t} y_2(t')dt'\right)^{k'}\right) y_2(t) \\ &= Z_{2,1}\{x(t)\} + Z_{2,2}\{y_2(t)\} + Z_{2,3}\{y_3(t)\}, \end{aligned} \tag{54}$$

and

$$\begin{aligned} &-\left(\sum_{k'=1}^{3} m_{3,k'}\left(-\int_{-\infty}^{t} y_3(t')dt'\right)^{k'}\right) y_3(t) \\ &= Z_{3,1}\{x(t)\} + Z_{3,2}\{y_2(t)\} + Z_{3,3}\{y_3(t)\}, \end{aligned} \tag{55}$$

where $y_r(t) = i_r(t)$ is approximated by the pth-order series expansion in Eq. (48). Equations (54)–(55) are the *dynamic equations* of each circuit from the class of Fig. 8. The next section explains how these equations may be used to determine analytical formulas in the spectral domain for the *Volterra kernels* of each of the two Volterra systems under analysis.

3.4 Analytical Derivation of Volterra Kernels

Application of the *harmonic probe technique* [15] to the dynamic equations (54)–(55) permits the derivation of the kernels in the spectral domain for each of the two Volterra systems under modelling. The idea behind this technique revolves around the application to Volterra system $r-1$ $r \in \{2,3\}$ of a m-tone complex exponential input, namely

$$x(t) = c \sum_{p'=1}^{m} x_{p'}(t), \tag{56}$$

with

$$x_{p'}(t) = \exp(j\omega_{p'} t), \tag{57}$$

to obtain the mth-order kernel $H_{r,m}(j\omega_1, \ldots, j\omega_m)$ ($c \in \mathbb{R}_+$ is a constant). Inserting Eq. (56) into Eq. (49), the nth-order component of the output of Volterra system $H_r[\cdot]$ in response to the aforementioned m-tone complex exponential input is

$$y_{r,n}(t) = H_{r,n}[x(t)] = c^n \sum_{p'_1=1}^{m} \cdots \sum_{p'_n=1}^{m} H_{r,n}(j\omega_{p'_1}, \ldots, j\omega_{p'_n}) e^{(j\omega_{p'_1}+\ldots+j\omega_{p'_n})t}. \tag{58}$$

At this point the series expansions for input, as expressed by Eq. (56) with $x_{p'}(t)$ defined in (57), and output, as given in Eq. (48) with $y_{r,n}(t)$ reported in (58), are inserted into the dynamic equations (54)–(55) of the circuit class under focus. Not all terms of the input and output series expansions need to be taken into account since computations within such equations are only allowed to deliver results containing factor c^m. Furthermore, within each of the resulting dynamic equations, only expressions proportional to $e^{(j\omega_1+j\omega_2+\ldots+j\omega_m)t}$ are retained, since only the mth-order kernel transform multiplied to $e^{(j\omega_1+j\omega_2+\ldots+j\omega_m)t}$ has distinct arguments $j\omega_1, j\omega_2, \ldots, j\omega_m$. The final equation is then solved for $H_{r,m}(j\omega_1, j\omega_2, \ldots, j\omega_m)$. The solution is found to depend upon the kernels[3] of order $n < m$. In order to derive the simplest possible expression for the mth-order kernel of Volterra system $r-1$ ($r \in \{2,3\}$), without loss of generality we assume that $H_{r,n}(j\omega_1, j\omega_2, \ldots, j\omega_n)$, i.e. the kernel of order n ($n \in \{1, 2, \ldots, m\}$) of Volterra system $r-1$ is symmetric [10]. In other words the order of appearance of variables $j\omega_1, j\omega_2, \ldots, j\omega_n$ within the argument of $H_{r,n}(\cdot)$ is

[3]For each Volterra system $H_r[\cdot]$, the harmonic probe technique requires the determination of all the kernels $H_{r,n}(j\omega_1, \ldots, j\omega_n)$ of order $n < m$ before $H_{r,m}(j\omega_1, \ldots, j\omega_m)$ may be derived.

irrelevant. Since any Volterra operator of order n may be equivalently characterised either by any of its n non-symmetric kernels or by its symmetrized kernel [10], this assumption is by no means restrictive. In the next section a closed form expression for the 1st-order kernel of Volterra system $r-1$ ($r \in \{2, 3\}$) is obtained.

3.4.1 Kernels of Order 1

With $m = 1$, Eqs. (56)–(57) give $x(t) = ce^{j\omega_1 t}$. The dynamic equations (54)–(55) of the circuit class under focus are allowed to contain terms with factor c only. Therefore, with reference to such equations, the left hand side is set to 0, while in the right hand side we replace $x(t)$ with $ce^{j\omega_1 t}$, and $y_r(t)$ with

$$y_{r,1}(t) = c^1 H_{r,1}(j\omega_1)e^{j\omega_1 t}, \tag{59}$$

as derived from Eq. (58). The dynamic equations (54)–(55) reduce to

$$0 = Z_{2,1}\{ce^{j\omega_1 t}\} + Z_{2,2}\{c^1 H_{2,1}(j\omega_1)e^{j\omega_1 t}\} + Z_{2,3}\{c^1 H_{3,1}(j\omega_1)e^{j\omega_1 t}\}, \tag{60}$$
$$0 = Z_{3,1}\{ce^{j\omega_1 t}\} + Z_{3,2}\{c^1 H_{2,1}(j\omega_1)e^{j\omega_1 t}\} + Z_{3,3}\{c^1 H_{3,1}(j\omega_1)e^{j\omega_1 t}\}. \tag{61}$$

From properties of integro-differential operators applied to complex exponential signals, we have

$$Z_{k,l}\{e^{j\omega t}\} = Z_{k,l}(j\omega)e^{j\omega t}, \tag{62}$$

for $k, l \in \{1, 2, 3\}$, where $Z_{k,l}(j\omega)$ denotes the element at row k and column l of the impedance matrix $\mathbf{Z}(j\omega)$, arranged as

$$\mathbf{Z}(j\omega) = \begin{pmatrix} Z_{1,1}(j\omega) & Z_{1,2}(j\omega) & Z_{1,3}(j\omega) \\ Z_{2,1}(j\omega) & Z_{2,2}(j\omega) & Z_{2,3}(j\omega) \\ Z_{3,1}(j\omega) & Z_{3,2}(j\omega) & Z_{3,3}(j\omega) \end{pmatrix}. \tag{63}$$

Solving equations (60)–(61) for $H_{2,1}(j\omega_1)$ and $H_{3,1}(j\omega_1)$ the following analytical expressions define the kernels of order 1 of the two Volterra systems under modeling:

$$H_{2,1}(j\omega_1) = \frac{Z_{2,1}(j\omega_1)Z_{3,3}(j\omega_1) - Z_{2,3}(j\omega_1)Z_{3,1}(j\omega_1)}{Z_{3,2}(j\omega_1)Z_{2,3}(j\omega_1) - Z_{2,2}(j\omega_1)Z_{3,3}(j\omega_1)}, \tag{64}$$
$$H_{3,1}(j\omega_1) = \frac{Z_{3,1}(j\omega_1)Z_{2,2}(j\omega_1) - Z_{3,2}(j\omega_1)Z_{2,1}(j\omega_1)}{Z_{3,2}(j\omega_1)Z_{2,3}(j\omega_1) - Z_{2,2}(j\omega_1)Z_{3,3}(j\omega_1)}. \tag{65}$$

The first-order kernel of Volterra system $r-1$ ($r \in \{2, 3\}$) depends only upon properties of the two-port of Fig. 8 through the elements of the impedance matrix in Eq. (63).

The analytical expression for the 2nd-order kernel of Volterra system $r-1$ ($r \in \{2, 3\}$) is presented in the next section.

3.4.2 Kernels of Order 2

With $m = 2$, Eqs. (56)–(57) give $x(t) = c\left(e^{j\omega_1 t} + e^{j\omega_2 t}\right)$. Further, Eq. (58) yields

$$y_{r,1}(t) = c^1\left(H_{r,1}(j\omega_1)e^{j\omega_1 t} + H_{r,1}(j\omega_2)e^{j\omega_2 t}\right) \tag{66}$$

$$y_{r,2}(t) = c^2\left(H_{r,2}(j\omega_1, j\omega_1)e^{2j\omega_1 t} + H_{r,2}(j\omega_2, j\omega_2)\ e^{2j\omega_2 t} + 2H_{r,2}(j\omega_1, j\omega_2)\right.$$
$$\left. e^{j(\omega_1+\omega_2)t}\right) \tag{67}$$

Here the dynamic equations (54)–(55) reduce to

$$m_{2,1}\left(\int_{-\infty}^{t} y_{2,1}(t')dt'\right) y_{2,1}(t) = Z_{2,2}\{y_{2,2}(t)\} + Z_{2,3}\{y_{3,2}(t)\}, \tag{68}$$

$$m_{3,1}\left(\int_{-\infty}^{t} y_{3,1}(t')dt'\right) y_{3,1}(t) = Z_{3,2}\{y_{2,2}(t)\} + Z_{3,3}\{y_{3,2}(t)\}. \tag{69}$$

Inserting equations (66)–(67) into Eqs. (68)–(69), solving the integrals, and assuming the second-order kernel of Volterra system $r-1$ ($r \in \{2, 3\}$) to be symmetric [10], a number of algebraic manipulations lead to the following expression for $H_{r,2}(j\omega_1, j\omega_2)$ ($r \in \{2, 3\}$):

$$H_{2,2}(j\omega_1, j\omega_2) = \left(\frac{Z_{3,3}(j(\omega_s))m_{2,1}H_{2,1}(j\omega_1)H_{2,1}(j\omega_2)}{\eta(j(\omega_s))}\right.$$
$$\left. - \frac{Z_{2,3}(j(\omega_s))m_{3,1}H_{3,1}(j\omega_1)H_{3,1}(j\omega_2)}{\eta(j(\omega_s))}\right)\frac{1}{2}\left(\frac{1}{j\omega_1} + \frac{1}{j\omega_2}\right), \tag{70}$$

and

$$H_{3,2}(j\omega_1, j\omega_2) = \left(\frac{Z_{22}(j(\omega_s))m_{3,1}H_{3,1}(j\omega_1)H_{3,1}(j\omega_2)}{\eta(j(\omega_s))}\right.$$
$$\left. - \frac{Z_{32}(j(\omega_s))m_{2,1}H_{2,1}(j\omega_1)H_{2,1}(j\omega_2)}{\eta(j(\omega_s))}\right)\frac{1}{2}\left(\frac{1}{j\omega_1} + \frac{1}{j\omega_2}\right), \tag{71}$$

where $\eta(\cdot) \triangleq Z_{3,3}(\cdot)Z_{2,2}(\cdot) - Z_{23}(\cdot)Z_{32}(\cdot)$, and $\omega_s \triangleq \omega_1 + \omega_2$. Note that the spectral domain representation of the second-order kernel of system $H_r[\cdot]$ ($r \in \{2, 3\}$) depends upon properties of the three-port through the elements of the impedance matrix, as well as upon characteristics of both memristors in Fig. 8 through parameters $m_{2,1}$ and $m_{3,1}$.

3.4.3 Kernels of Order 3

For $m = 3$ Eqs. (56)–(57) give $x(t) = c\left(e^{j\omega_1 t} + e^{j\omega_2 t} + e^{j\omega_3 t}\right)$. Further, Eq. (58) yields

$$y_{r,1}(t) = c^1\Big(H_{r,1}(j\omega_1)e^{j\omega_1 t} + H_{r,1}(j\omega_2)e^{j\omega_2 t} + H_{r,1}(j\omega_3)e^{j\omega_3 t} \Big) \tag{72}$$

$$\begin{aligned} y_{r,2}(t) = c^2 \Big(& H_{r,2}(j\omega_1, j\omega_1)e^{2j\omega_1 t} + H_{r,2}(j\omega_2, j\omega_2)e^{2j\omega_2 t} + H_{r,2}(j\omega_3, j\omega_3)e^{2j\omega_3 t} \\ & +2H_{r,2}(j\omega_1, j\omega_2)e^{j(\omega_1+\omega_2)t} + 2H_{r,2}(j\omega_1, j\omega_3)e^{j(\omega_1+\omega_3)t} + 2e^{j(\omega_2+\omega_3)t} \\ & \cdot H_{r,2}(j\omega_2, j\omega_3) \Big) \end{aligned} \tag{73}$$

$$\begin{aligned} y_{r,3}(t) = c^3 \Big(& H_{r,3}(j\omega_1, j\omega_1, j\omega_1)e^{3j\omega_1 t} + H_{r,3}(j\omega_2, j\omega_2, j\omega_2)e^{3j\omega_2 t} + e^{3j\omega_3 t} \\ & \cdot H_{r,3}(j\omega_3, j\omega_3, j\omega_3) + 6H_{r,3}(j\omega_1, j\omega_2, j\omega_3)e^{j(\omega_1+\omega_2+\omega_3)t} + 3e^{j(2\omega_1+\omega_2)t} \\ & \cdot H_{r,3}(j\omega_1, j\omega_1, j\omega_2) + 3H_{r,3}(j\omega_1, j\omega_2, j\omega_2)e^{j(\omega_1+2\omega_2)t} + 3e^{j(2\omega_1+\omega_3)t} \\ & \cdot H_{r,3}(j\omega_1, j\omega_1, j\omega_3) + 3H_{r,3}(j\omega_1, j\omega_3, j\omega_3)e^{j(\omega_1+2\omega_3)t} + 3e^{j(2\omega_2+\omega_3)t} \\ & \cdot H_{r,3}(j\omega_2, j\omega_2, j\omega_3) + 3H_{r,3}(j\omega_2, j\omega_3, j\omega_3)e^{j(\omega_2+2\omega_3)t} \Big) \end{aligned} \tag{74}$$

Omitting the explicit dependence of all variables upon time to improve formatting, the dynamic equations (54)–(55) reduce to

$$\begin{aligned} \left(m_{2,1}\left(\int_{-\infty}^{t} y_{2,2}dt'\right) - m_{2,2}\left(\int_{-\infty}^{t} y_{2,1}dt'\right)^2 \right) & y_{2,1} + m_{2,1}\left(\int_{-\infty}^{t} y_{2,1}dt'\right) y_{2,2} \\ & = Z_{2,2}\{y_{2,3}\} + Z_{2,3}\{y_{3,3}\}, \end{aligned} \tag{75}$$

$$\begin{aligned} \left(m_{3,1}\left(\int_{-\infty}^{t} y_{3,2}dt'\right) - m_{3,2}\left(\int_{-\infty}^{t} y_{3,1}dt'\right)^2 \right) & y_{3,1} + m_{3,1}\left(\int_{-\infty}^{t} y_{3,1}dt'\right) y_{3,2} \\ & = Z_{3,2}\{y_{2,3}\} + Z_{3,3}\{y_{3,3}\}, \end{aligned} \tag{76}$$

Let us now insert Eqs. (72), (73), and (74) into Eqs. (75)–(76). Solving the integrals and taking into account the symmetric property of the third-order kernel of Volterra system $r-1$ ($r \in \{2, 3\}$) [10], the expression for $H_{r,3}(j\omega_1, j\omega_2, j\omega_3)$ ($r \in \{2, 3\}$) is found to be given by

$$\begin{aligned} H_{2,3}(j\omega_1, j\omega_2, j\omega_3) = & -\frac{m_{2,2}Z_{3,3}(j\tilde{\omega}_s)}{3\eta(j\tilde{\omega}_s)}\left(\prod_{i=1}^{3} H_{2,1}(j\omega_i)\right)\left(\frac{1}{j\omega_1 j\omega_2} + \frac{1}{j\omega_1 j\omega_3}\right. \\ & \left. + \frac{1}{j\omega_2 j\omega_3}\right) + \frac{m_{2,1}Z_{3,3}(j\tilde{\omega}_s)}{3\eta(j\tilde{\omega}_s)}\Big[H_{2,2}(j\omega_1, \omega_2)H_{2,1}(j\omega_3) \\ & \cdot \left(\frac{1}{j\omega_3} + \frac{1}{j(\omega_1+\omega_2)}\right) + H_{2,2}(j\omega_1, \omega_3)H_{2,1}(j\omega_2)\left(\frac{1}{j\omega_2}\right. \end{aligned}$$

$$
\begin{aligned}
&+\frac{1}{j(\omega_1+\omega_3)}\Bigg)+H_{2,2}(j\omega_2,\omega_3)H_{2,1}(j\omega_1)\left(\frac{1}{j(\omega_2+\omega_3)}\right.\\
&\left.+\frac{1}{j\omega_1}\right)\Bigg]-\Bigg[H_{3,2}(j\omega_1,\omega_2)H_{3,1}(j\omega_3)\left(\frac{1}{j(\omega_1+\omega_2)}\right.\\
&\left.+\frac{1}{j\omega_3}\right)+H_{3,2}(j\omega_1,\omega_3)H_{3,1}(j\omega_2)\left(\frac{1}{j\omega_2}+\frac{1}{j(\omega_1+\omega_3)}\right)\\
&+H_{3,2}(j\omega_2,\omega_3)H_{3,1}(j\omega_1)\left(\frac{1}{j\omega_1}+\frac{1}{j(\omega_2+\omega_3)}\right)\Bigg]\frac{m_{3,1}}{3}\\
&\cdot\frac{Z_{2,3}(j\tilde{\omega}_s)}{\eta(j\tilde{\omega}_s)}+\frac{m_{3,2}Z_{2,3}(j\tilde{\omega}_s)}{3\eta(j\tilde{\omega}_s)}\left(\prod_{i=1}^{3}H_{3,1}(j\omega_i)\right)\left(\frac{1}{j\omega_1 j\omega_2}\right.\\
&\left.+\frac{1}{j\omega_1 j\omega_3}+\frac{1}{j\omega_2 j\omega_3}\right)
\end{aligned}
\tag{77}
$$

and

$$
\begin{aligned}
H_{3,3}(j\omega_1,j\omega_2,j\omega_3)=&-\frac{m_{3,2}Z_{2,2}(j\tilde{\omega}_s)}{3\eta(j\tilde{\omega}_s)}\left(\prod_{i=1}^{3}H_{3,1}(j\omega_i)\right)\left(\frac{1}{j\omega_1 j\omega_2}+\frac{1}{j\omega_1 j\omega_3}\right.\\
&\left.+\frac{1}{j\omega_2 j\omega_3}\right)+\frac{m_{3,1}Z_{2,2}(j\tilde{\omega}_s)}{3\eta(j\tilde{\omega}_s)}\Bigg[H_{3,2}(j\omega_1,\omega_2)H_{3,1}(j\omega_3)\\
&\left(\frac{1}{j\omega_3}+\frac{1}{j(\omega_1+\omega_2)}\right)+H_{3,2}(j\omega_1,\omega_3)H_{3,1}(j\omega_2)\left(\frac{1}{j\omega_2}\right.\\
&\left.+\frac{1}{j(\omega_1+\omega_3)}\right)+H_{3,2}(j\omega_2,\omega_3)H_{3,1}(j\omega_1)\left(\frac{1}{j(\omega_2+\omega_3)}\right.\\
&\left.+\frac{1}{j\omega_1}\right)\Bigg]-\Bigg[H_{2,2}(j\omega_1,\omega_2)H_{2,1}(j\omega_3)\left(\frac{1}{j(\omega_1+\omega_2)}\right.\\
&\left.+\frac{1}{j\omega_3}\right)+H_{2,2}(j\omega_1,\omega_3)H_{2,1}(j\omega_2)\left(\frac{1}{j\omega_2}+\frac{1}{j(\omega_1+\omega_3)}\right)\\
&+H_{2,2}(j\omega_2,\omega_3)H_{2,1}(j\omega_1)\left(\frac{1}{j\omega_1}+\frac{1}{j(\omega_2+\omega_3)}\right)\Bigg]\frac{m_{2,1}}{3}\\
&\frac{Z_{3,2}(j\tilde{\omega}_s)}{\eta(j\tilde{\omega}_s)}+\frac{m_{2,2}Z_{3,2}(j\tilde{\omega}_s)}{3\eta(j\tilde{\omega}_s)}\left(\prod_{i=1}^{3}H_{2,1}(j\omega_i)\right)\left(\frac{1}{j\omega_1 j\omega_2}\right.\\
&\left.+\frac{1}{j\omega_1 j\omega_3}+\frac{1}{j\omega_2 j\omega_3}\right)
\end{aligned}
\tag{78}
$$

where $\eta(\cdot)$ has the same definition as in Sect. 3.4.2, i.e. $\eta(\cdot) \triangleq Z_{3,3}(\cdot)Z_{2,2}(\cdot) - Z_{23}(\cdot)Z_{32}(\cdot)$, and $\tilde{\omega}_s \triangleq \omega_1 + \omega_2 + \omega_3$. Note that $H_{r,3}(j\omega_1, j\omega_2, j\omega_3)$ is more sensitive to the characteristics of both memristors of Fig. 8 as compared to the expression for $H_{r,2}(j\omega_1, j\omega_2)$ ($r \in \{2, 3\}$). This may be inferred by comparing Eq. (70) with Eq. (77) and Eq. (71) with Eq. (78).

For each Volterra system under modelling kernels of higher order may be determined in a similar manner. Given the knowledge of the first p elements from the kernel set of Volterra system $H_r[\cdot]$, its response to any excitation may be approximated by a simple algebraic sum, as specified by Eqs. (48)–(49). This bypasses any inconvenience associated with the numerical integration of the ordinary differential equations describing the two-memristor circuit under study. In order to further clarify this point, in the next section the analytical expression for the output of Volterra system $r-1$ under a sine-wave input is given in terms of its kernels in the spectral domain.

3.5 Response to a Sine-Wave Input

Under a certain input $x(t)$, knowledge of all kernels up to the pth order allows the computation of the pth-order approximation $y_r^{(p)}(t)$ of the response $y_r(t)$ of Volterra system $r-1$ through the use of Eqs. (48)–(49). In these equations the kernels appear in the time domain, while in Sects. 3.4.1, 3.4.2, and 3.4.3 they were derived in the spectral domain. The mth-order kernel of Volterra system $r-1$ in the time-domain, i.e. $h_{m,r}(\tau_1, \tau_2, \ldots, \tau_m)$, may be obtained from its mathematical description in the spectral domain, i.e. $H_m(j\omega_1, j\omega_2, \ldots, j\omega_m)$, using [10]

$$h_{r,m}(\tau_1, \ldots \tau_m) = \int_{-\infty}^{+\infty} \ldots \int_{-\infty}^{+\infty} \frac{H_{r,m}(j\omega_1, \ldots, j\omega_m)}{(2\pi)^m} e^{(j\omega_1\tau_1 + \cdots + j\omega_m\tau_m)} d\omega_1 \ldots d\omega_m. \quad (79)$$

Nonetheless, there exists an alternative approach allowing the direct use of the kernels in the spectral domain for the computation of the approximated system response $y_r^{(p)}(t)$ in Eq. (48). Let us show how. Without loss of generality, let us assume that the input to Volterra system $r-1$ is a sine wave expressed by $x(t) = x_0 \sin(\omega_0 t) = \frac{1}{2j} \sum_{i=-1, i\neq 0}^{+1} x_i \text{sign}(i) e^{j\omega_i t}$ where $x_i = x_0$ for $i \in \{-1, 1\}$, $\omega_1 = -\omega_{-1} = \omega_0$, and sign$(\cdot)$ stands for the signum function. Inserting this into Eq. (49), and manipulating the resulting expression, the nth-order Volterra series of $y_r^{(p)}(t)$, i.e. $y_{r,n}(t)$, is found to be directly related to the spectral domain representation of the nth-order Volterra kernel [16], assuming the following form:

$$y_{r,n}(t) = H_{r,n}[x(t)] = \sum_{i_1=-1,\ i_1\neq 0}^{+1} \cdots \sum_{i_n=-1,\ i_n\neq 0}^{+1} H_{r,n}(j\omega_{i_1}, \ldots j\omega_{i_n}) e^{j(\omega_{i_1}+\ldots+\omega_{i_n})t} \frac{1}{(2j)^n} x_{i_1} \text{sign}(i_1) \ldots x_{i_n} \text{sign}(i_n) \quad (80)$$

Using equality

$$H_{r,1}(-j\omega_0)e^{-j\omega_0 t} = \left(H_{r,1}(j\omega_0)e^{j\omega_0 t}\right)^*,$$

for $n = 1$ Eq. (80) delivers

$$y_{r,1}(t) = x_0 \Im\{H_{r,1}(j\omega_0)e^{j\omega_0 t}\}. \tag{81}$$

Similarly, employing relations

$$H_{r,2}(-j\omega_0, -j\omega_0)e^{-2j\omega_0 t} = \left(H_{r,2}(j\omega_0, j\omega_0)e^{2j\omega_0 t}\right)^*,$$
$$H_{r,2}(j\omega_0, -j\omega_0) = \left(H_{r,2}(-j\omega_0, +j\omega_0)\right)^*$$

and, given the symmetric property of the second-order kernels, adopting the equivalence

$$\Re\{H_{r,2}(j\omega_0, -j\omega_0)\} = H_{r,2}(j\omega_0, -j\omega_0),$$

for $n = 2$ Eq. (80) assumes the following form:

$$y_{r,2}(t) = \frac{x_0^2}{2}\left(H_{r,2}(j\omega_0, -j\omega_0) - \Re\{H_{r,2}(j\omega_0, j\omega_0)e^{2j\omega_0 t}\}\right). \tag{82}$$

Finally, setting $n = 3$, taking into account the symmetric property of the third-order kernels, and making use of equation

$$H_{r,3}(-j\omega_0, -j\omega_0, -j\omega_0)e^{-3j\omega_0 t} = \left(H_{r,3}(j\omega_0, j\omega_0, j\omega_0)e^{3j\omega_0 t}\right)^*,$$

Equation (80) yields

$$y_{r,3}(t) = -\frac{x_0^3}{4}\Im\{H_{r,3}(j\omega_0, j\omega_0, j\omega_0)e^{3j\omega_0 t}\} + \frac{3x_0^3}{4}\Im\{H_{r,3}(j\omega_0, j\omega_0, -j\omega_0)e^{j\omega_0 t}\}. \tag{83}$$

Higher-order Volterra operators may be derived in a similar manner from Eq. (80). Finally, the sum of all Volterra operators $H_{r,n}[\cdot]$ for $n \in \{1, \ldots, p\}$, according to Eq. (48), provides the desired expression for the pth-order approximation $y_r^{(p)}(t)$ of the response $y_r(t)$ of Volterra system $r - 1$ to the sine wave input $x(t) = x_0 \sin(\omega_0 t)$.

The next Section presents a case study of an exemplifying circuit from the class of Fig. 8 to demonstrate the applicability of the Volterra series approach to the modelling of the two-memristor circuits under study.

3.6 Theory Validation

Figure 10 shows an oscillator within the class of two-memristor circuits of Fig. 8.

The dotted rectangular box identifies a three-port characterized by an impedance matrix – see Eq. (63) – with elements given by

Table 3 Parameter setting for the simulation of the circuit of Fig. 10

$R/k\Omega$	C/μF	L/mH
10	2	50
$m_{r,0}/\Omega$	$m_{r,1}/(\Omega\, C^{-1})$	$\mathbf{z}(0)$
5	500	$\mathbf{0}$

$$\begin{aligned}
Z_{11}(j\omega) &= Z_{13}(j\omega) = Z_{31}(j\omega) = R,\\
Z_{12}(j\omega) &= Z_{21}(j\omega) = 0,\\
Z_{22}(j\omega) &= j\omega L + m_0,\\
Z_{23}(j\omega) &= Z_{32}(j\omega) = -j\omega L, \quad \text{and}\\
Z_{33}(j\omega) &= R + m_0 + j\omega L + \frac{1}{j\omega C}.
\end{aligned}$$

Defining the states of the electronic system of Fig. 10 as $z_1 \triangleq v_C$, $z_2 \triangleq i_L$, $z_3 \triangleq q_2 = \int_{-\infty}^{t} i_2(t')dt'$ and $z_4 \triangleq q_3 = \int_{-\infty}^{t} i_3(t')dt'$, where v_C is the voltage across the capacitor, i_L the current through the inductor, and q_r the charge at port r ($r \in \{2, 3\}$), applying circuit-theoretic principles [17] to this two-memristor circuit, the nonlinear ODE governing its dynamical behaviour may be cast as

$$\begin{aligned}
\dot{z}_1 &= \frac{1}{C}\frac{Ri_1(t) - z_1 + (m_0 + M(-z_3))\, z_2}{R + 2m_0 + M(-z_3) + M(-z_4)},\\
\dot{z}_2 &= \frac{m_0 + M(-z_3)}{L}\frac{Ri_1(t) - z_1 - (R + m_0 + M(-z_4))z_2}{R + 2m_0 + M(-z_3) + M(-z_4)},\\
\dot{z}_3 &= -\frac{Ri_1(t) - z_1 - (R + m_0 + M(-z_4))\, z_2}{R + 2m_0 + M(-z_3) + M(-z_4)}, \quad \text{and}\\
\dot{z}_4 &= -\frac{Ri_1(t) - z_1 + (m_0 + M(-z_3))\, z_2}{R + 2m_0 + M(-z_3) + M(-z_4)}.
\end{aligned} \tag{84}$$

A sine-wave current $i_1(t) = i_0 \sin(\omega t) = x_0 \sin(\omega t)$ is now inserted at port 1. The aim is to compare the results of the numerical integration of the ODE set (84) with the Volterra series-based model prediction.

The input amplitude and angular frequency are respectively set to $x_0 = 0.4\, A$, and $\omega = 20\, rads^{-1}$.

The values for resistor, capacitor, and inductor as well as the non-null coefficients $m_{r,0}$ and $m_{r,1}$ in the memristance function $M(q_{m,r})$ of each memristor m_r ($r \in \{2, 3\}$), and the initial condition $\mathbf{z}(0) = [z_1(0), z_2(0), z_3(0), z_4(0)]'$ for the numerical integration of equation (84) are reported in Table 3.

In the Volterra series-based model of the circuit of Fig. 10, the output $y_r(t)$ of the $(r-1)$th Volterra system $H_r[\cdot]$ ($r \in \{2, 3\}$) is approximated by expanding the Volterra series $y_r^{(p)}(t)$ in Eq. (48) up to the 3rd-order, i.e. p is set to 3.

The time waveform of $q_2(t)$ ($q_3(t)$) resulting from numerical integration of the ODE set (84) – red curve – and from the Volterra series (VS) expansion derived by

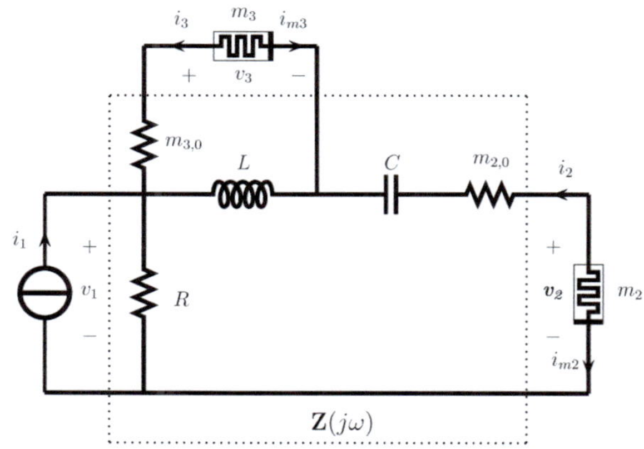

Fig. 10 Example of a two-memristor circuit from the class of Fig. 8

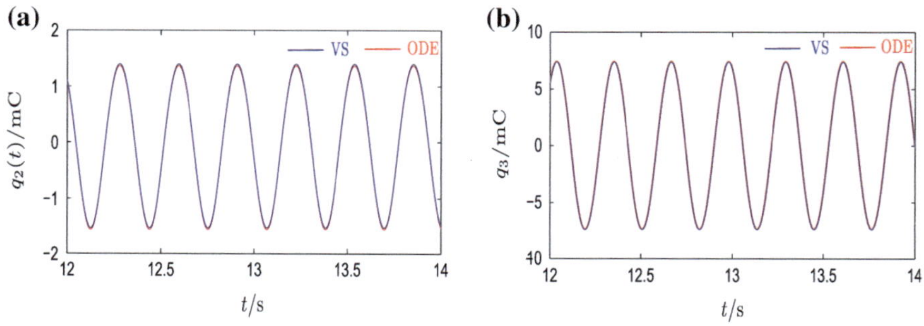

Fig. 11 Charges $q_2(t)$ at port 2 (**a**) and $q_3(t)$ at port 3 (**b**) over time, resulting from numerical integration of (84) (red curve) and Volterra series-based analysis (blue curve)

taking the time integral of the output $y_2^{(3)}$ ($y_3^{(3)}$) of Volterra system $H_2[\cdot]$ ($H_3[\cdot]$) – blue curve – is shown in Fig. 11a, b.

Fig. 12a, b show the current-voltage loci of memristors m_2 and m_3, respectively. Red and blue curves refer to numerical and Volterra series-based solutions, respectively. In both figures the voltage $v_{m,r}$ across memristor m_r ($r \in \{2, 3\}$) is calculated from Eq. (44). Therefore it is the sum between the voltage at port r and the drop across linear resistor $m_{r,0}$. Here m_2 (see Fig. 12a), and m_3 (see Fig. 12b) exhibit the typical fingerprint of memristors, i.e. a pinched hysteresis loop under sinusoidal excitation [1].

The power spectra of current $i_{m,2}(t)$ ($i_{m,3}(t)$) at port 2 (3) is shown in Fig. 13a, b. Numerical integration results and Volterra-based model predictions (the same colour coding employed in the previous figures applies here) are in good agreement.

The good agreement between Volterra theory-based model predictions and ODE numerical solutions for the illustrative circuit of Fig. 10 demonstrates the applicability of the proposed approach to the mathematical description of two-memristor circuits from the class in Fig. 8.

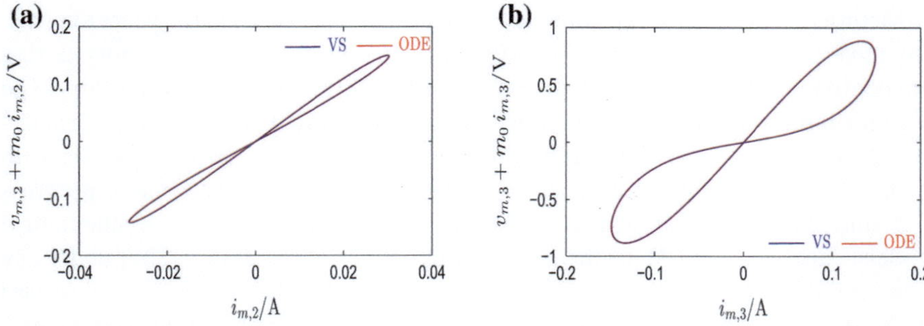

Fig. 12 Current-voltage loci for memristor m_2 (**a**) and m_3 (**b**). Red curve: numerical simulation result. Blue curve: Volterra series-based mathematical description

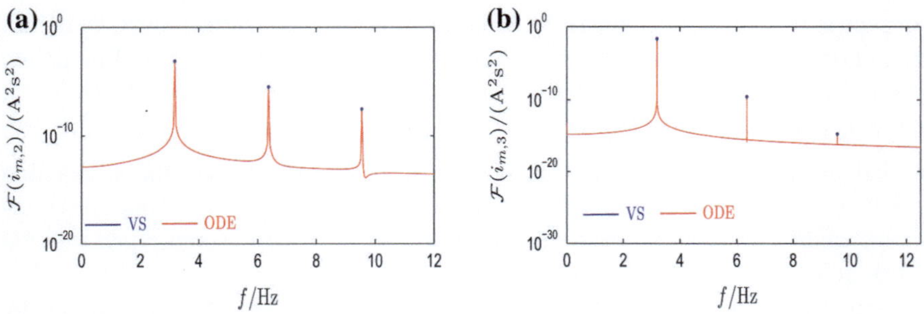

Fig. 13 Power spectra of currents i_2 at port 2 (**a**) and i_3 at port 3 (**b**). Numerical results and Volterra series representations are shown in red and blue, respectively

In conclusion, the proposed systematic modeling technique may be easily generalized to multi-memristor circuits, consisting of linear and dynamic m'-ports with one-port driven by a current source and the remaining $m' - 1$ loaded by ideal generic current-controlled memristors.

Further, applying the principle of duality [18], similar guidelines as reported here may allow the analytical representation of linear and dynamic m'-ports with one port excited by a voltage source and the remaining $m' - 1$ closed on ideal generic voltage-controlled memristors.

4 Conclusions

Industries [19] and academic researchers [11, 12] have shown an exponentially growing interest in memristors and their applications [20] since the first conscious observation [2] of memristor fingerprints [21] in a nanoscale titanium oxide-based resistance switching memory [3].

Memristors promise to revolutionize the world of electronics in the years to come. The most popular application fields of memristors are nonvolatile memory design and neuromorphic system development. However, this is only part of the story. The inherent nonlinear dynamics of memristors may be exploited to design novel dynamical circuits processing signals differently from conventional electronic systems, with the aim to extend and/or improve their functionalities [9]. To foster major progress in memristor circuit and system design, the development of better mathematical models, capable to reproduce the nonlinear dynamics of the devices [22] under any input/initial condition combination, is one of the top priorities for research. Standard modelling approaches may be inadequate to describe the rich nonlinear dynamics of memristors. Nonlinear system theory may provide alternative strategies. In this respect the Volterra series paradigm [10] constitutes a suitable theoretical framework for the modelling of memristor devices, circuits and systems [23]. As a proof-of-concept in this work we applied the Volterra theory to the mathematical characterization of a class of ideal generic memristor [11] circuits. The proposed modeling technique is based upon a systematic procedure consisting of a limited number of mathematical steps, listed below.

- First the circuit under focus needs to be inspected to identify the association between system theoretic variables and electrical circuit's physical quantities (currents and/or voltages) for each sub-circuit whereby a Volterra series representation is required.
- Then the dynamic equations of the circuit under consideration are derived by applying standard concepts from nonlinear circuit theory [17].
- Subsequently a rigorous mathematical analysis of the circuit response to a multitone excitation [15] allows the derivation of analytical formulas for the Volterra kernels of each of the aforementioned sub-circuits in the spectral domain.
- The response of any of the Volterra systems, representing the aforementioned sub-circuits, to any input of interest is finally obtained directly from the sum of multi dimensional convolution integrals involving input and kernels.

In order to reduce computing complexity, only a finite number of kernels are typically determined. As a result, the Volterra series modeling of these sub-circuits will be inevitably characterized by a certain degree of accuracy.[4]

Each element from the class of circuits under modelling in these lecture notes consists of a linear and dynamic m'-port with one port excited by a current source and all remaining $m' - 1$ ports closed on an ideal generic memristor [11] For simplicity of presentation, this work analysed the cases $m' = 2$ and $m' = 3$ only, but the proposed modelling technique applies to larger values for m' as well. In order to demonstrate the validity of the theoretical approach, illustrative circuits from the aforementioned class were analyzed. By applying Kirchhoff's laws, ordinary differential equations governing the time evolution of all voltages and currents in these

[4]Increasing the number of terms in the Volterra series expansion, the proposed system theoretic representation may attain any desired degree of accuracy. The derivation of a larger number of kernels may be necessary in case the nonlinearity of the circuit under study is high.

circuits were determined. The numerical integration of the resulting equations agreed well with the mathematical description based upon the theory of Volterra.

This study proves that the Volterra series paradigm may be adopted to model ideal generic memristors and circuits based upon them. In the future the proposed system theoretic technique will be applied to generic and extended memristor devices and circuits.

In conclusion, it is well known that circuit- and system-theoretic concepts may considerably support the work of circuit designers [24]. Particularly, the design optimization phase of a nonlinear circuit may be facilitated upon identification of its representative Volterra kernel set. In fact, under these conditions, the derivation of an optimal circuit parameter set may involve just a couple of steps. Analytical expressions for performance measures of interest – e.g. the total harmonic distortion or the signal-to-noise ratio – in terms of the kernels (and, consequently, of the circuit parameters) would be derived first. Then the application of a standard optimization procedure aimed at minimizing a certain cost function, defined in terms of the aforementioned performance measures, would deliver an optimal circuit parameter set. Using this approach the design optimization task would be accomplished faster [25] than would be the case in standard circuit simulation-based iterative procedures.

Acknowledgements The authors would like to acknowledge the contribution of the EU COST Action IC1401 MemoCiS. The Deutsche Forschung Gesellschaft (DFG) for their financial contribution to the research project "Locally active memristive data processing (LAMP)" under grant number TE 257/22-1 is thanked.
This work has been partially supported by the Czech Science Foundation under grant No. 14 – 19865S, Czech Republic.

References

1. Chua, L.O.: Memristor: the missing circuit element. IEEE Trans. Circuit Theory **18**(5), 507–519 (1971)
2. Strukov, D.B., Snider, G.S., Stewart, D.R., Williams, R.S.: The missing memristor found. Nat. Lett. **453**, 80–83 (2008). https://doi.org/10.1038/nature06932
3. Chua, L.O.: Resistance switching memories are memristors. Appl. Phys. A **102**, 765–783 (2011)
4. Ascoli, A., Corinto, F., Senger, V., Tetzlaff, R.: Memristor model comparison. IEEE Circuits Syst. Mag. **13**(2), 89–105 (2013). https://doi.org/10.1109/MCAS.2013.2256272
5. Ascoli, A., Tetzlaff, R., Chua, L.O.: Robust simulation of a TaO memristor model. Radioengineering **24**(2), 384–392 (2015). https://doi.org/10.13164/re.2015.0001
6. Corinto, F., Ascoli, A., Gilli, M.: Nonlinear dynamics of memristor oscillators. IEEE Trans. Circuits Syst.–I **58**(6), 1323–1336 (2011). https://doi.org/10.1109/TCSI.2010.2097731
7. Corinto, F., Ascoli, A., Gilli, M.: Analysis of current-voltage characteristics for memristive elements in pattern recognition systems. Int. J. Circuit Theory Appl. **40**(12), 1277–1320 (2012). https://doi.org/10.1002/cta.1804
8. Ascoli, A., Corinto, F.: Memristor models in chaotic neural circuits. Int. J. Bifurc. Chaos Appl. Sci. Eng. **23**(3), 1350052(28) (2013). https://doi.org/10.1142/S0218127413500521
9. Ascoli, A., Slesazeck, S., Mähne, H., Tetzlaff, R., Mikolajick, T.: Nonlinear dynamics of a locally-active memristor. IEEE Trans. Circuits Syst.–I: Regul. Pap. **62**(4), 1165–1174 (2015). https://doi.org/10.1109/TCSI.2015.2413152

10. Schetzen, M.: The Volterra and Wiener Theories of Nonlinear Systems. Wiley, New York (1980)
11. Chua, L.O.: Everything you wish to know about memristors but are afraid to ask. RadioEngineering **24**(2), 319–368 (2015)
12. Chua, L.O.: If it's pinched, it's a memristor. Semicond. Sci. Technol. **29**, 104001 (2014). https://doi.org/10.1088/0268-1242/29/10/104001
13. Chua, L.O., Ng, C.Y.: Frequency domain analysis of nonlinear systems. IEE J. Electron. Circuits Syst. **3**(4), 165–185 (1979)
14. Ascoli, A., Schmidt, T., Tetzlaff, R., Corinto, F.: Application of the Volterra series paradigm to memristive systems. In: Tetzlaff, R. (ed.) Memristors and Memristive Systems, pp. 163–191. Springer, New York (2014)
15. Boyd, S., Tang, Y.S., Chua, L.O.: Measuring Volterra kernels. IEEE Trans. Circuits Syst. **CAS–30**(8), 571–577 (1983)
16. Lang, Z.-Q., Billings, S.A.: Evaluation of output frequency responses of nonlinear systems under multiple inputs. IEEE Trans. Circuits Syst.-II **47**(1), 28–38 (2000)
17. Chua, L.O., Desoer, C.A., Kuh, E.A.: Linear and Nonlinear Circuits. McGraw-Hill, New York (1985)
18. Chua, L.O.: Introduction to Nonlinear System Theory. McGraw-Hill, New York (1969)
19. https://www.micron.com/about/emerging-technologies/3d-xpoint-technology
20. Ascoli, A., Tetzlaff, R., Corinto, F.: A class of versatile circuits, made up of standard electrical components, are memristors. Int. J. Circuit Theory Appl. **44**, 127–146 (2015). https://doi.org/10.1002/cta.2067
21. Adhikari, S.P., Sah, M.P., Kim, H., Chua, L.O.: Three fingerprints of memristor. IEEE Trans. Circuits Syst.–I Reg. Pap. **60**(11), 3008–3021 (2013)
22. Ascoli, A., Corinto, F., Tetzlaff, R.: Generalized boundary condition memristor model. Int. J. Circuit Theory Appl. **44**, 60–84 (2015). https://doi.org/10.1002/cta.2063
23. Ascoli, A., Tetzlaff, R., Chua, L.O.: Everything you wish to know about memristors, but are afraid to ask. Plenary Session IEEE Nano, Rome, Italy (2015)
24. Ascoli, A., Curran, P., Feely, O.: Modeling the dynamics of log-domain circuits. Int. J. Circuit Theory Appl. **35**, 33–70 (2007). https://doi.org/10.1002/cta.379
25. Bauer, A., Schwarz, W.: Circuit analysis and optimization with automatically derived Volterra kernels. Int. Symp. Circuits Syst. **I**, 491–494 (2000). (Geneve)

A Memristor-Based Cell for Complexity

Arturo Buscarino, Claudia Corradino, Luigi Fortuna, Mattia Frasca and Viet-Thanh Pham

Abstract The theoretical importance of memristor goes much beyond the field, i.e., circuit theory, in which its discovery originated. In fact, neuroscience and nonlinear science in general are also interested as memristors have been found to be adequate model of both synapses and Hodgkin–Huxley axons. In this work, we discuss the use of memristors as the unique nonlinear element of the cell of a cellular architecture reproducing several phenomena of interest for nonlinear science such as autowave propagation and Turing pattern formation. We illustrate the model and present numerical simulations showing how the same cell structure can account for these different dynamical behaviors when its parameters are varied.

1 Introduction

The memristor is considered one of the four basic circuit elements along with the resistor, the inductor and the capacitor. The history of its discovery and recent interest it is attracting is nowadays quite well-known: despite its existence was postulated by Leon O. Chua in 1971 [1], its discovery is more recent as experimental findings on a device with a memristive behavior were performed in the Hewlett-Packard

A. Buscarino · C. Corradino · L. Fortuna · M. Frasca (✉)
Dipartimento di Ingegneria Elettrica Elettronica e Informatica,
University of Catania, viale A. Doria 6, 95125 Catania, Italy
e-mail: mattia.frasca@dieei.unict.it

A. Buscarino
e-mail: arturo.buscarino@dieei.unict.it

C. Corradino
e-mail: claudia.corradino@dieei.unict.it

L. Fortuna
e-mail: luigi.fortuna@dieei.unict.it

V.-T. Pham
School of Electronics and Telecommunications, Hanoi University
of Science and Technology, 01 Dai Co Viet, Hanoi, Vietnam
e-mail: pvt3010@gmail.com

F. Corinto and A. Torcini (eds.), *Nonlinear Dynamics in Computational Neuroscience*, PoliTO Springer Series, https://doi.org/10.1007/978-3-319-71048-8_8

laboratories in 2008 [2]. After this seminal paper [2], it was realized [3, 4] that many other devices (e.g., thin films with resistance switching properties), known and used even before 2008, are memristors.

The memristor is a two-terminal electronic device displaying a functional relationship between the time integral of the current through it and that of the voltage across its terminals. This device is now gaining considerable interest for its potential applications in the design of logic gates [5, 6], nonvolatile memories [7], and neuromorphic systems [8]. In addition, memristor proves to be an element of extreme importance in complexity science and engineering. For instance, it has been widely used as a key component to build periodic and chaotic oscillators [9–11] as well as in more complex structures made of many interacting cells [12–14]. In the context of complexity science and, in particular, in neuroscience, of fundamental importance are also the results of Ref. [15] showing that the Hodgkin–Huxley axon is modeled by memristors.

Our work collocates in the framework of complex systems and focuses on the use of memristors as the fundamental nonlinearity of the basic unit of a reaction-diffusion system. In particular, we show that memristor-based reaction-diffusion circuits can be used to generate nonlinear complex phenomena such as autowaves and Turing patterns. The first ones are spatio-temporal phenomena of great importance in neuron dynamics, while Turing patterns are steady-state heterogeneous solutions fundamental for the study of morphogenesis [16]. We employ a cellular nonlinear network (CNN) with memristors as the architecture of our system and demonstrate the simplicity (in terms of the number of elements comprised in the basic cell) and the emergent features deriving from the presence of the memristors.

2 The Memristor-Based Cell

The elementary cell considered in this work is a circuit with three components, an inductor, a capacitor, and an active memristor, connected in a parallel configuration as shown in Fig. 1. The active memristor may be viewed as a single device or a negative conductance in parallel with a passive memristor.

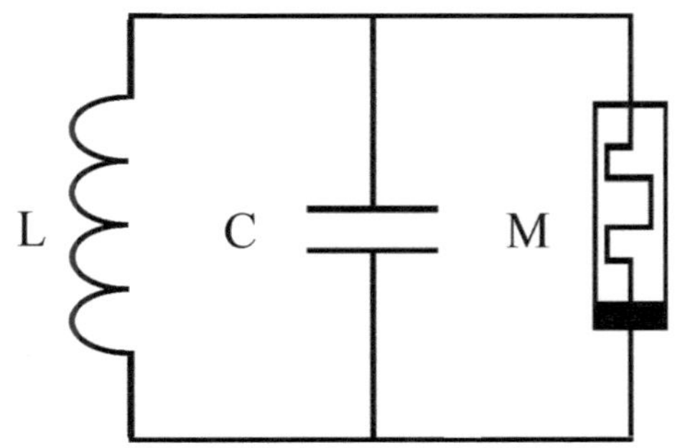

Fig. 1 The basic cell

By applying Kirchhoff's circuit laws and the constitutive equations of the memristor, the dynamics of the cell is derived:

$$\begin{cases} C\dot{v} = -i - W(\varphi)\, v + Gv, \\ L\dot{i} = v, \\ \dot{\varphi} = v, \end{cases} \tag{1}$$

where v, i, and φ represents the voltage across the capacitor, the current in the inductor, and the memristor flux, respectively. The memristor is described by

$$W(\varphi) = \frac{dq(\varphi)}{d\varphi} = \begin{cases} a & |\varphi| < 1, \\ b & |\varphi| > 1, \end{cases} \tag{2}$$

with $q(\varphi)$ being the following piecewise–linear function:

$$q(\varphi) = b\varphi + 0.5\,(a - b)\,(|\varphi + 1| - |\varphi - 1|)\,. \tag{3}$$

Model (1) is rewritten in terms of dimensionless variables with the following positions: $x = v$, $y = i$, $z = \varphi$, $\alpha = 1/C$, $\beta = 1/L$, and $\gamma = G$, so that one obtains:

$$\begin{cases} \dot{x} = \alpha\,(-y - W(z)\,x + \gamma x)\,, \\ \dot{y} = \beta x, \\ \dot{z} = x, \end{cases} \tag{4}$$

Note that the last two equations of (4) are not independent:

$$\dot{y} - \beta\dot{z} = 0 \tag{5}$$

Integrating Equation (5) and solving for z, one gets

$$z = \frac{y}{\beta} + c, \tag{6}$$

where c is a constant term that depends on the memristor initial conditions. Substituting the relationship (6) into Eq. (4) we finally obtain

$$\begin{aligned} \dot{x} &= \alpha(-y + W(\tfrac{y}{\beta} + c)x) \\ \dot{y} &= \beta x \end{aligned} \tag{7}$$

representing the dimensionless equations of the basic cell. Equation (7) are functions of several parameters and of the memristor initial conditions. It turns out that changing them different dynamical behaviors can be obtained. In particular, we are interested to two possibilities. The first one is the possibility to generate oscillations with a slow-fast dynamics. An example of such behavior is illustrated in Fig. 2 for $\alpha = 10$, $\gamma = 0.3$, $\beta = 0.01$, $a = 0.1$, $b = 0.5$ and $c = 0$. We note that, in the slow

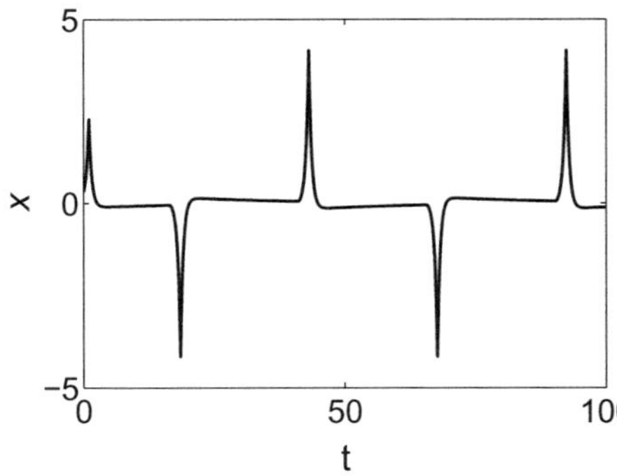

Fig. 2 Oscillations exhibited by the cell (7) when the parameters are fixed as $\alpha = 10$, $\gamma = 0.3$, $\beta = 0.01$, $a = 0.1$, $b = 0.5$ and $c = 0$. The oscillations are characterized by slow-fast dynamics

part of the trajectory, the state variable remains at an almost constant value for a considerably long period of time τ_{st}. After this long period, the state has a large excursion and rapidly returns to the previous state in a significantly short period of time τ_{ex}, where $\tau_{ex} << \tau_{st}$. Similar slow-fast dynamics is peculiar of many biological systems (neurons for instance) and is a prerequisite for the generation of autowaves [17, 18].

The second dynamical behavior we are interested in is connected to the conditions for the generation of Turing patterns. A reaction-diffusion equation can produce Turing patterns if some conditions at the level of the microscopic unit composing the system are satisfied. In particular, the cell should admit an equilibrium that is stable when the cell is isolated and becomes unstable due to the effect of the diffusion (diffusion-driven instability).

The cell (7) admits an equilibrium point in the origin. Linearizing the dynamics around this equilibrium and studying its stability properties through the analysis of the Jacobian matrix, we derive that

$$\begin{array}{l}\alpha W(c) < 0\\ \alpha\beta > 0\end{array} \tag{8}$$

must be satisfied to obtain that the equilibrium point is stable. A further set of conditions derive from imposing the diffusion-driven destabilization of the equilibrium point and will be, therefore, discussed after introducing the complete reaction-diffusion system.

3 Model of the Memristive Cellular Nonlinear Network

Autowaves and Turing patterns are nonlinear phenomena both appearing in reaction-diffusion equations. We illustrate here the reaction-diffusion system based on the memristive cell described in the previous Section. We will refer to such architecture as a reaction-diffusion memristive cellular nonlinear network (MCNN).

Consider a generic reaction-diffusion model with continuous-time second order dynamics for each microscopic unit and discrete space:

$$\begin{aligned} \dot{x}_{i,j} &= f(x_{i,j}, y_{i,j}) + D_{11}\nabla^2 x_{i,j} + D_{12}\nabla^2 y_{i,j} \\ \dot{y}_{i,j} &= g(x_{i,j}, y_{i,j}) + D_{21}\nabla^2 x_{i,j} + D_{22}\nabla^2 y_{i,j} \end{aligned} \tag{9}$$

where the discrete-space Laplacian is given by:

$$\begin{aligned} \nabla^2 x_{i,j} &= x_{i+1,j} + x_{i-1,j} + x_{i,j+1} + x_{i,j-1} - 4x_{i,j} \\ \nabla^2 y_{i,j} &= y_{i+1,j} + y_{i-1,j} + y_{i,j+1} + y_{i,j-1} - 4y_{i,j} \end{aligned} \tag{10}$$

being i and j the spatial indices ($i = 1, \ldots, M$ and $j = 1, \ldots, N$). In Eq. (9) $\dot{x}_{i,j} = f(x_{i,j}, y_{i,j})$ and $\dot{y}_{i,j} = g(x_{i,j}, y_{i,j})$ represent the reactive part, while $D_{11}\nabla^2 x_{i,j} + D_{12}\nabla^2 y_{i,j}$ and $D_{21}\nabla^2 x_{i,j} + D_{22}\nabla^2 y_{i,j}$ the diffusive part that includes terms acting on the same layer, i.e., *self-diffusive* terms ($D_{11}\nabla^2 x_{i,j}$ and $D_{22}\nabla^2 y_{i,j}$) and terms acting on the other layer, i.e., *cross-diffusive* terms ($D_{12}\nabla^2 y_{i,j}$ and $D_{21}\nabla^2 x_{i,j}$) (Fig. 3).

The system under our investigation is built starting from the reaction-diffusion model (9) and using the dynamics of the memristive cell in Fig. 1 as reactive part of the model, that is, considering $f(x_{i,j}, y_{i,j}) = \alpha(-y_{i,j} + W(\frac{y_{i,j}}{\beta} + c)x_{i,j})$ and $g(x_{i,j}, y_{i,j}) = \beta x_{i,j}$. We obtain a grid of $M \times N$ cells described by the following equations:

$$\begin{aligned} \dot{x}_{i,j} &= \alpha(-y_{i,j} + W(\tfrac{y_{i,j}}{\beta} + c)x_{i,j}) + D_{11}\nabla^2 x_{i,j} + D_{12}\nabla^2 y_{i,j} \\ \dot{y}_{i,j} &= \beta x_{i,j} + D_{21}\nabla^2 x_{i,j} + D_{22}\nabla^2 y_{i,j} \end{aligned} \tag{11}$$

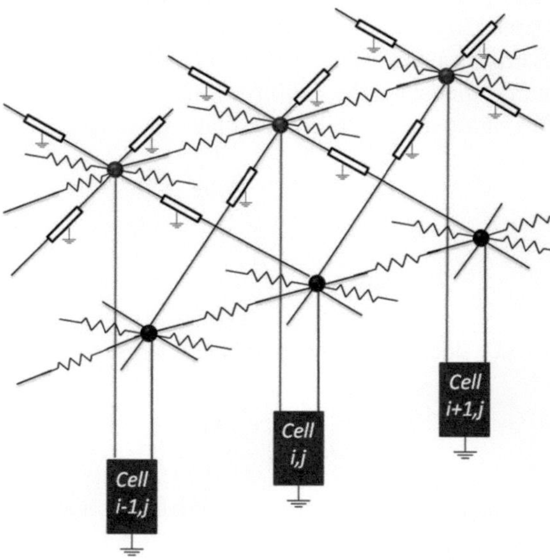

Fig. 3 Schematic representation of a reaction-diffusion MCNN with both self and cross diffusion. Resistors implements self-diffusion coefficients D_{11} and D_{22}, while the two-port blocks realize the cross-diffusion terms

In the next Section, we will show that, while to generate autowaves self-diffusion on the first layer is enough, both self- and cross-diffusion are required to obtain Turing patterns in the structure.

4 Generation of Complex Phenomena

The term *autowaves* or *autonomous waves* was introduced to model phenomena appearing in combustion, phase transitions, chemical reactions, and many biological processes and characterized by waves having amplitude and shape that do not change during propagation, not exhibiting reflection or interference, but annihilating when two wavefronts collide [17]. Autowaves are obtained in the MCNN when the single unit has slow-fast dynamics and a suitable value of the self-diffusion coefficient in the first layer, that is D_{11}, is set (the other diffusion terms may be set to zero) [19].

Consider, for instance, a 50×50 structure with zero-flux boundary conditions, $D_{11} = 0.51$, $D_{12} = D_{21} = D_{22} = 0$, $\alpha = 10$, $\gamma = 0.3$, $\beta = 0.01$, $a = 0.1$, $b = 0.5$ and $c = 0$. The initial conditions are selected as $x_{i,j}(0) = y_{i,j}(0) = 0$, where $1 \leq i, j \leq 50$, except $x_{25,2}(0) = x_{25,49}(0) = x_{26,2}(0) = x_{26,49}(0) = 1.5$, so that there are two seeds of wave excitation, one located in the left and the other in the right of the MCNN.

The result is the formation of two circular wavefronts generated at opposite sides of the structure and propagating with the same velocity in opposite directions. Fig. 4 illustrates several snapshots of this spatio-temporal phenomenon. When the two waves collide ($t = 12$), they finally annihilate each other rather than penetrating one another, and, therefore, no interference takes place.

The same structure, using different parameters, is also able to generate Turing patterns [20]. In this case, the parameters for the single units have to be selected so that Eq. (8) are satisfied. These conditions alone are not sufficient as we have seen that they guarantee the stability of the equilibrium point for the isolated cell, but not diffusion-driven instability. This latter is satisfied if, along with (8), it holds that:

$$
\begin{array}{l}
D_{22}\alpha W(c) + D_{21}\alpha - D_{12}\beta > 0 \\
(D_{22}\alpha W(c) + D_{21}\alpha - D_{12}\beta)^2 - 4\det(D)\alpha\beta > 0 \\
\det(\mathrm{D}) = D_{11}D_{22} - D_{12}D_{21} > 0 \\
\det(\mathrm{D}_s) > 0
\end{array}
\tag{12}
$$

where D_s denotes the symmetrical part of the matrix

$$
\mathrm{D} = \begin{bmatrix} D_{11} & D_{12} \\ D_{21} & D_{22} \end{bmatrix}
\tag{13}
$$

These further conditions are derived by applying a widely used approach for reaction-diffusion systems based on CNNs, that is, the technique of *spatial eigenfunction-based decoupling* [21, 22], and further imposing dissipativity of the

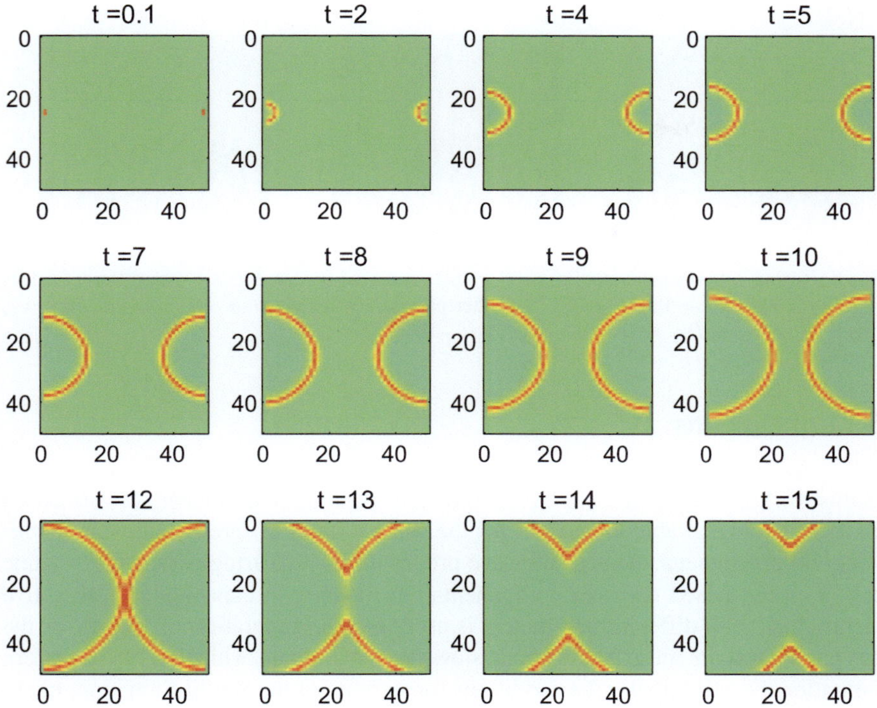

Fig. 4 Propagation and annihilation of autowaves in a reaction-diffusion MCNN with $D_{11} = 0.51$ and $D_{12} = D_{21} = D_{22} = 0$. The other parameters are fixed as in Fig. 2

diffusion terms. Note that the conditions for Turing patterns formation, Eqs. (8) and (12), cannot be satisfied if cross-diffusion is not considered, i.e., $D_{12} = D_{21} = 0$.

Let us now consider a set of parameters satisfying conditions (8) and (12), namely, $\alpha = 1$, $\beta = 0.8$, $D_{11} = 1$, $D_{12} = 4$, $D_{21} = 6$, $D_{22} = 26$, $a = -0.01$, $b = -0.61$, and three different values of c ($c = 0$, $c = 0.2$, and $c = 0.5$) and let us consider a 100×100 reaction-diffusion MCNN with zero-flux boundary conditions and initial conditions, without loss of generality, randomly selected with uniform distribution in $[0, 1]$. Turing patterns are generated in all the three cases (see Fig. 5, showing layer x) with a remarkable property: the final pattern observed depends on the value of c. Varying c controls the type of patterns: from spots to stripes and mixed spot/stripe pattern. As the parameter c accounts for the effect of the initial conditions of the memristor, we found that these have a strong influence on the final global behavior of the structure. Since all the other parameters are not changed, we are in presence of a bifurcation without parameter, a preeminent feature of circuits based on memristors [23].

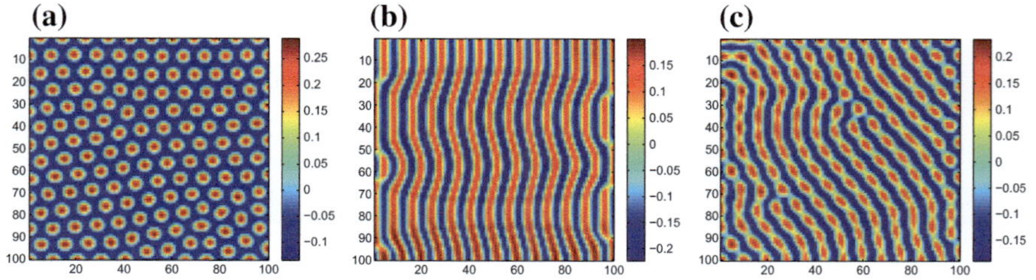

Fig. 5 Turing patterns generated in a 100 × 100 reaction-diffusion MCNN with different values of c: **a** $c = 0.5$; **b** $c = 0$; **c** $c = 0.2$. The other parameters are set as: $\alpha = 1$, $\beta = 0.8$, $D_{11} = 1$, $D_{12} = 4$, $D_{21} = 6$, $D_{22} = 26$, $a = -0.01$, $b = -0.61$

5 Conclusions

In this work we have discussed a reaction-diffusion cellular nonlinear network based on memristors and able to account for the generation of different nonlinear spatio-temporal phenomena such as autowave propagation and Turing patterns. The same cell, a circuit based on three components (an inductor, a capacitor and an active memristor), with different parameters is used. For the generation of autowaves the cell parameters are set so that to obtain slow-fast oscillations, while for Turing pattern formation the cell satisfies the conditions for stability of the equilibrium point of the isolated dynamics and diffusion-driven instability. It is remarkable that, in terms of the number of required components, this cell is particularly simple.

Another interesting feature that the reaction-diffusion MCNN exhibits is the presence of bifurcations without parameters. This appears evident from the analysis of the Turing patterns generated by the structure. Different types of patterns are obtained when the initial conditions of the memristors are changed, while the other structural parameters of the system are kept fixed. Using memristors in a reaction-diffusion model, thus, allows a great simplification of the basic cell design and provides a further degree of freedom controlling the dynamics of the spatio-temporal phenomena occurring in the structure.

References

1. Chua, L.O.: Memristor-the missing circuit element. IEEE Trans. Circuit Theory **18**(5), 507–519 (1971)
2. Strukov, D.B., Snider, G.S., Stewart, D.R., Williams, R.S.: The missing memristor found. Nature **453**, 80–83 (2008)
3. Chua, L.O.: Resistance switching memories are memristors. Appl. Phys. A **102**(4), 765–783 (2011)
4. Chua, L.O.: Everything you wish to know about memristors but are afraid to ask. Radioengineering **24**(2), 319–368 (2015)

5. Borghetti, J., Snider, G.D., Kuekes, P.J., Yang, J.J., Stewart, D.R., Williams, R.S.: Memristive switches enable stateful logic operations via material implication. Nature **464**, 876–878 (2010)
6. Vourkas, I., Papandroulidakis, G., Vasileiadis, N., Sirakoulis, G.: Boolean logic operations and computing circuits based on memristors. IEEE Trans. Circuits Syst.-II **61**, 972–976 (2014)
7. Ho, Y., Huang, G.M., Li, P.: Dynamical properties and design analysis for nonvolatile memristor memories. IEEE Trans. Circuits Syst.-I **58**, 724–736 (2011)
8. Jo, S.H., Chang, T., Ebong, I., Bhadviya, B.B., Mazumder, P., Lu, W.: Nanoscale memristor device as synapse in neuromorphic systems. Nano Lett. **10**, 1297–1301 (2010)
9. Itoh, M., Chua, L.O.: Memristor oscillators. Int. J. Bifurc. Chaos **18**(11), 3183–3206 (2008)
10. Corinto, F., Ascoli, A., Gilli, M.: Nonlinear dynamics of memristor oscillators. IEEE Trans. Circuits Syst. I **58**(6), 1323–1336 (2011)
11. Buscarino, A., Fortuna, L., Frasca, M., Gambuzza, L.V.: A chaotic circuit based on Hewlett-Packard memristor. Chaos **22**(2), (2012)
12. Duan, S., Hu, X., Dong, Z., Wang, L., Mazumder, P.: Memristor-based cellular nonlinear/neural network: design, analysis, and applications. IEEE Trans. Neural Netw. Learn. Syst. **26**(6), 1202–13 (2015)
13. Kim, H., Sah, M.P., Yang, C., Roska, T., Chua, L.O.: Memristor bridge synapses. Proc. IEEE **100**(6), 2061–2070 (2012)
14. Gambuzza, L.V., Buscarino, A., Fortuna, L., Frasca, M.: Memristor-based adaptive coupling for consensus and synchronization. IEEE Trans. Circuits Syst. I: Regul. Pap. **62**(4), 1175–1184 (2015)
15. Chua, L.O., Sbitnev, V., Kim, H.: Hodgkin-Huxley axon is made of memristors. Int. J. Bifurc. Chaos **22**(3), 1230011 (2012)
16. Murray, J.D.: Mathematical Biology I: An Introduction. Springer, New York (2002)
17. Perez-Munuzuri, V., Perez-Villar, V., Chua, L.O.: Autowaves for image processing on a two-dimensional CNN array of excitable nonlinear circuits: flat and wrinkled labyrinths. IEEE Trans. Circuits Syst. I: Fundam. Theory Appl. **40**(3), 174–181 (1993)
18. Arena, P., Caponetto, R., Fortuna, L., Manganaro, G.: Cellular neural networks to explore complexity. Soft Comput. **1**(3), 120–136 (1997)
19. Pham, V.T., Buscarino, A., Fortuna, L., Frasca, M.: Autowaves in memristive cellular neural networks. Int. J. Bifurc. Chaos **22**(8), 1230027 (2012)
20. Buscarino, A., Corradino, C., Fortuna, L., Frasca, M., Chua, L.O.: Turing patterns in memristive cellular nonlinear networks. IEEE Trans. Circuits Syst. I: Regul. Pap. **63**(8), 1222–1230 (2016)
21. Goras, L., Chua, L.O., Leenaerts, D.M.W.: Turing patterns in CNNs. I. once over lightly. IEEE Trans. Circuits Syst. I: Fundam. Theory Appl. **42**(10), 602–611 (1995)
22. Goras, L., Chua, L.O.: Turing patterns in CNNs. II. equations and behaviors. IEEE Trans. Circuits Syst. I: Fundam. Theory Appl. **42**(10), 612–626 (1995)
23. Corinto, F., Forti, M.: Memristor circuits: flux-charge analysis method. IEEE Trans. Circuits Syst. I: Regul. Pap. (2016) (in press)

MIX
Papier aus verantwortungsvollen Quellen
Paper from responsible sources
FSC® C105338

If you have any concerns about our products,
you can contact us on
ProductSafety@springernature.com

In case Publisher is established outside the EU,
the EU authorized representative is:
Springer Nature Customer Service Center GmbH
Europaplatz 3, 69115 Heidelberg, Germany

Printed by Libri Plureos GmbH
in Hamburg, Germany